North Bridge

Ship Yard

Ship Yard

RIVER KEEL

Garrison Street

Side

Fishersgate

Quaker
Meeting

Hadby Road

...on Square

Workhouse

HARBOUR

Harbour Side

Wellington Street

Magazine

Citadel

Soldiers' Barracks

Officers' Quarters

Middlethorpe
1847

GREAT ESTUARY

A Natural History

A NOVEL

Keith Oatley

VIKING

VIKING

Published by the Penguin Group

Penguin Books Canada Ltd, 10 Alcorn Avenue, Toronto, Ontario, Canada M4V 3B2

Penguin Books Ltd, 27 Wrights Lane, London w8 5TZ, England

Penguin Putnam Inc., 375 Hudson Street, New York, New York 10014, U.S.A.

Penguin Books Australia Ltd, Ringwood, Victoria, Australia

Penguin Books (NZ) Ltd, cnr Rosedale and Airborne Roads, Albany, Auckland 1310, New Zealand

Penguin Books Ltd, Registered Offices: Harmondsworth, Middlesex, England

First published 1998

10 9 8 7 6 5 4 3 2 1

Publisher's note: This book is a work of fiction. Apart from those indicated in the Acknowledgements at the back of the book, names, characters, places and incidents either are the product of the author's imagination or are used fictitiously, and any resemblance to actual persons living or dead, events, or locales is entirely coincidental.

Printed and bound in Canada on acid free paper ∞

CANADIAN CATALOGUING IN PUBLICATION DATA

Oatley, Keith
A natural history

ISBN 0-670-88167-8

I. Title

PS8579.A75N37 1998   C813'.54   C98-930937-1
PR9199.3.O27N37 1998

Visit Penguin Canada's web site at **www.penguin.ca**

Dedicated to my Toronto friends

Contents

It is perfectly true, as philosophers say, that life must be understood backwards. But they forget the other proposition that it must be lived forwards. And if one thinks over that proposition it becomes more and more evident that life can never really be understood in time simply because at no particular moment can I find the necessary resting place from which to understand it—backwards.

Søren Kierkegaard, *Journals*, 1843

9. Two are better than one; because they have a good reward for their labour.
10. For if they fall, the one will lift up his fellow: but woe to him that is alone when he falleth; for he hath not another to help him up.
11. Again, if two lie together, then they have heat; but how can one be warm alone?

Ecclesiastes, IV

A Natural History

I.
Fuga

Practice

... a person in the best of health when he is smitten by the cholera, is in an instant transformed into a corpse! There is the same state of the eyes, the same aspect of the face; the same coldness of the limbs; the same colour of the skin, etc. Were not the intelligence left, so to speak, intact, were there not a remnant of the voice, even though almost imperceptible, one would proceed to burial at the actual moment of the attack.

Letter of 3ʳᵈ December 1831 from the physiologist François Magendie during his inspections of the cholera in Sunderland, to the President of the Academy of Sciences in Paris.

———◦———

Chapter One

If I could understand this one disease, thought Leggate, everything would become clear.

He peered at his map of Middlethorpe, drew a small circle and then, in a careful hand, wrote within it, "282." He laid down his pen, straightened himself, and gazed across the town's miniature expanse. He examined again his most recent clump of circles: 279, 280, 281, 282. Each marked a death from cholera.

Death, that most elemental of human losses, so impossible to countenance, was reduced to numbers so that we rather than it might prevail.

Leggate took a long breath, sat up, linked his fingers and stretched his arms upwards. He rose and opened his window, increasing the noise from the lane below. When he had taken over the practice there had been a garden opposite, but soon afterwards it was replaced by a coal yard. On the second storey he still had a view of the overcrowded river thirty yards away. A laden ship manoeuvred, pouring indiscriminate smoke from a funnel with two gold bands around it. He picked up a brass telescope, and pointed it towards the ship.

Is this how it comes, he thought, from Hamburg, ashore along mooring lines like the rats, perhaps with the rats themselves? There must be some substance, a poison, from abroad, or does it lurk here always, waiting to be released?

He focused his telescope on the ship's paddle-wheel as it agitated the brown water that lay calm, the incoming tide having reached its height, the water near the tops of sea walls and quays. The paddle-wheel stopped, reversed, churned the water, and with it a swirl of floating straw and pig-sty rubbish.

Disgusting, he thought, filthy straw and excreta, all thrown into the river like so much else, all mixed together every day, then up and down with the tide.

He collapsed the telescope with his palms, pulled the window shut, returned to his map: here the town's sprawl of streets, here the harbour office, infirmary, theatre, the principal church, upriver the new waterworks and baths.

On tracing paper he had drawn lines of contour to represent not the heights of hills but the earnings of the people who lived there. He had made an obsessive table of wages: shipping clerks, lightermen, dockers, candle-makers, all the rest… He could overlay the contour lines on the map in which buildings and streets—other facts of human life—were represented.

The next wave of cholera is gathering, he thought. It'll come ashore somewhere here, or if not here, it'll be here soon after, once the wave forms it advances, inexorable.

He rose and went to lie on his couch. He closed his eyes to remember himself stooping into an ill-lit room, and his first sight of cholera. His nose wrinkled once more at that odour of vomit mixed with some more repel-lent smell. A man lay on a low bed, damp with excretions, his body shrunken, the face's blue skin clung in folds to bony contours, accentuating the skull beneath. A wife stood silent in a corner. The man turned his head a fraction towards Magendie, who put his fingers to the man's carotid artery, then after a few seconds, said: "*Regarde ça, Leggate.*"

Magendie spoke in French, but in the habit of hospital doctors who demonstrate to students, it would have been no different if the patient spoke his language; the convention was that Latin terms made medical talk opaque except to the initiated. "See this," said Magendie. "Pulse absent. Cadaveriza-tion, that's what this disease does, turns the body immediately into a corpse."

Leggate had been assistant to Professor Magendie, and had travelled with his master from Paris to the English port of Sunderland to peer at the dis-tending shadow of cholera when it spread across Europe in 1831. He thought again of the wife who saw her husband defeated in a day, without even being able to keep him clean; she and her children were for the workhouse.

Now sixteen years later he pursued the task that the visit to Sunderland had begun.

Breaking apart those whom God has joined, he thought, as he remem-bered again the woman standing mute at the side of the room. Her loss could call up his own echo of separation. Is it our Fallen state? he wondered. Shall it always be thus? We must break the power of disease. Magendie's observations were acute, but such wild inferences, too eager for some astonishing conclusion—cadaverization indeed! Approaching from this other direction I shall track it down, and give something to the world.

He had adopted many of Magendie's attitudes, not all his conclusions. Science instilled a confidence in self, an obligation to disagree where necessary. It had marked him like the scar of inoculation against smallpox. But a pockmark is superficial. Inoculation, though it protects against one disease, is itself another with workings that continue within. His inward disease was an obsession with cholera, a passion that could exclude all else—almost like falling in love.

He rose from his couch and looked again at the large map. He looked along the row of notebooks above his writing table: events captured from the flux of life. They bore ink-written legends: *Diary*, each with a date. Then notebooks of *Papers and Monographs Read*. Next were *Cholera, Propagation and Physiology, Vol. I* and *Vol. II*, then *Cholera Cases*, then *Daybook of Patients Visited*, and *Accounts*.

Cholera Cases had been started in 1845 when he had come to Middlethorpe. Corresponding to each circled number on his map was a paragraph in this book, describing a death from the disease. The town employed a registrar whose records included addresses of those who died when the epidemic struck in 1832. He visited each address, and aimed to know each victim's occupation, symptoms, details of others in the house who had contracted the disease but not succumbed. There were lacunae with the lapse of years, but allowing for these, he had succeeded. Now, in 1847, only a few dozen more little circles were needed to represent all those of Middlethorpe and its adjacent parishes who had died.

My notes accumulate, he thought. Positive knowledge, this time my article will be applauded. There are patterns, concentrations of death among the poor, and I can show that more accurately than anyone, but it's not enough, I can show more, so the pattern will make sense—this agglomeration of circles here, for instance, what does that mean? Go again, with mind receptive, and talk to the wives...

Wives, he thought. When I marry she'll be proud that I've freed the world of the disease, of course we'd take a better house, though this one's not unhealthy, no one here caught it. The doctor I bought it from attended the victims, but what exempts us doctors but not relatives, not women who lay out bodies?

He walked towards the fireplace, rested his elbow on the mantelpiece, and imagined himself addressed by the President of the Royal Society.

"How, Dr Leggate, would you describe the principal feature of your maps?"

"The principal feature, sir, is that deaths are unevenly distributed. If every tenth person had died, circles on my maps would be homogeneous in proportion to the population in each area. Or, if the disease attacked the indigent only, the deaths would be spread evenly among them. But this is not the pattern. In the workhouse, among the most impoverished, there was only one fatality. Therefore there must be something else present in some areas, but not others."

"What is it about the densely affected areas?"

"There may be disease-bearing miasmata, I cannot say with certainty, not as yet."

Not yet ... impressions, he thought. A right impression will come, accumulation of foetid air, or damp. But the idea of miasma, that the filth and stench of particular places equal disease, seems all wrong. Some places are morbid, but there's filth all the time, and the cholera certainly comes from abroad. So, from first principles, first define where it struck most fiercely, then from the where, I shall perceive the what. It will crack open, and I'll see it.

"Right!" he said, out loud to himself. "And then ..."

He was aware of making such utterances, meaningless, but somehow comforting to emphasize the progression of the inner argument that constantly filled his mind: "And then ..." he said out loud again, though nothing followed the "then."

He lay down once more on his couch, imagined the announcement in *The Times*: "Dr John Leggate solves riddle of the cause of cholera."

Footsteps sounded on the stairs; there was a knock and the face of the housekeeper appeared round the door.

"I hope you haven't forgotten Dr Newcombe's patient you promised to see at four o'clock, sir, you remember?"

He felt a sudden lurch of the stomach with the realization that he had forgotten. "Thank you, Mrs Huggins. Thank you."

"It's the Rector of Swanscombe, in Sutcliffe Street. Here's his name and address, I've written them down for you again."

Chapter Two

The sound of a piano: Marian was playing one of Mendelssohn's "Songs Without Words" when its yearning melody was jangled faintly by the doorbell. She was expecting her friend Caroline. She was at the open door of her music room before the bell had stopped. She ran downstairs, went ahead of the maid who made her way more slowly up from the kitchen, opened the front door. The next thing she then looked upon was a gentleman.

The outside light was bright. Marian blinked once. The gentleman removed his hat, bowed slightly.

"Dr Leggate," he said. "Standing in for Dr Newcombe, to see Mr Brooks."

She saw a clean-shaven face, held his gaze for an instant.

"It's my father. He's still quite unwell," she said. "The maid will show you in."

Inside, Marian Brooks went upstairs again, disappointed her friend had not arrived. She picked out a transcription: Mendelssohn again, "Overture to *A Midsummer Night's Dream*." Halfway through a phrase she rose, walked to her front window. She stood beside an easel on which stood an oil portrait of a handsome fair-haired young woman, and looked down to the street.

Patterns of sound in the air, she thought, I throw them up and they hang there, but ephemeral, just for a moment . . . That doctor has an educated accent, wonder why I've not met him before—is he new? but the smile? Doctors are usually more aloof.

She walked to the fireplace, extended her own face towards the pier-glass. Brown eyes set above high cheekbones gazed back. Friendly, she thought, a friendly face—she considered friendliness her chief characteristic. She saw chestnut hair pinned up loosely; it had slipped into an unruly shape, a bird's nest. She touched it with her hand, then left it with a shrug.

She thought again of the doctor introducing himself. He seems so . . . eager! Is Caroline late? Not yet, I suppose.

"To see Mr Brooks"—she spoke out loud, imitating his accent. What was his name? she thought.

She walked to her door, crossed the landing, ran lightly down the stairs once more, not stopping in the hall, but continuing down the back stairs to the kitchen.

"Is Mrs Brooks still out?"

"She'll be back after five, miss," said the maid.

Marian ascended to the hall, knocked once on a mahogany door, heard a faint invitation, entered her parents' bedroom. Her father was sitting up in bed, nightshirt rucked at his neck. The doctor on the far side of the bed, bending towards him, was gentle, like nothing so much as a mother with a baby.

Mr Brooks's posture prompted Marian to a pang of compassion. He too can be vulnerable, she thought. But this room's sweltering with that huge fire, can that be good for a fever?

The doctor listened at one end of a trumpet-shaped object, pressing the other end to his patient's back. He moved the device from place to place, said, "Breathe in, breathe out, in . . . and again, please."

Marian examined this doctor, appraised his thinnish face, dark tousled hair, the complexion not quite English, Mediterranean perhaps, and with the slight convex angle to the nose that she found touching. She imagined modelling it in clay, imagined moving her forefinger downwards with utmost slowness to make the shallow groove from nose to upper lip.

The doctor straightened, still regarding his patient who sat up and sought to arrange himself into some more proper posture. Marian waited by the door.

These illnesses, she thought, brought on by cold weather, fogs from the estuary, his lungs not strong, or is it some weakness of the throat? Caroline thinks it's because he's angry with my brother, and with not being able to move about because of his rheumatism. But with him it's always the same: sore throat, irritable seeking for sympathy, fever, then it passes.

The doctor drew a chair to the side of the bed, then said to his patient, "I think the disease is at its peak. Your throat's inflamed, and the lymph nodes are swollen. These cause the soreness, and make your neck tender, but the appearance is different from putrid fever, and you have no pneumonia. Best to keep in bed, sleep, have your daughter read to you."

"Very well, thank you," Mr Brooks growled with painful voice. Turning away from the doctor, he drew his knees up to his chest, cuddling the bed-clothes towards him, as if to sleep.

Leggate glanced up, the first sign that he had noticed Marian there. "Miss Brooks," he said. He rose from the chair, took up his bag from the floor, put it on the bed, took out some notebooks, another book, a handsome writing case, put them all on the bed.

I thought they were supposed to carry their instruments and medicines in their bags, she thought. All this one has is books.

"Make sure your father drinks a lot. Fresh water's best," said Leggate as he stowed his stethoscope in the bag, then the things he had taken out. "But he can have tea, or whatever he prefers. Keep the body's fluids up." Only at this point did he meet her eyes. "Let's hope he is on the mend."

"Can you say nothing certain?" she asked.

He showed no sign of irritability. He stood upright to face her. He was taller than she had thought when he came in, with an inner energy. "Despite descriptions of fevers for two thousand years," he said, "almost nothing is known about how they work."

"But shouldn't you act on the disease?"

"To act effectively one must understand what one acts upon. Most remedies act only on the credulity of the patient."

Leggate smiled, giving off a warmth that reassured Marian, although she felt provoked by his remark. Her hand rose to touch her hair. Had it slipped farther?

"No letting blood to draw off the poison?" she asked.

"Certainly not blood." He sounded incredulous. He walked to the end of the bed, took two steps towards her, looking at her as if she were a fellow medical student. "Utter nonsense," he said, "like throwing salt over your shoulder." He made an illustrative cast with his hand. "Worse really"— now he was excited. "At least throwing salt does no harm! There were ex-periments in Paris with letting blood in cases of pneumonia, and what d'you think?"

"To test the effect, you mean?"

"No advantage whatever to patients relieved of blood. In fact, if they recovered, they took longer to do so."

She was taken aback. "Why is this not generally known?"

"The medical profession doesn't like it, but most remedies lack any rational foundation."

"I suppose some artifice is usual when money is to change hands." She looked at him slyly.

"Earlier you reproached me for refusing to employ a useless remedy."

"*Touché*," she said. "So I was wrong to do so. See how I yield to reason?" But she was not giving up. "You mean I shouldn't chastise you, but your profession?"

"A few of us try to apply scientific principles."

"You are a scientist?"

"We're just on the edge of understanding how the body works. The body's a machine. When we do understand it, most of these medical tricks and potions and whatnot will be seen for what they are . . . hogwash."

Surprised by this farmyard expression, Marian said, "So science decrees that we remain passive."

"Not passive; doing nothing's probably the most important curative principle we have." He grinned to disarm her. "The body has profound healing powers, far more effective than anything we can do. Our job is to assist them."

Then, as if they would together hatch a plan, he said: "But if you like, we will act. Why not some Battley's Liquor? It won't speed recovery—but it contains opium which will make for rest, and soothe the cough a bit."

"Now we learn that some remedies are effective?"

"I shall write the prescription." The moment was past. The doctor's speech ran once more along rails of medical convention. "I'll let Dr Newcombe know I've called."

From his bag he took his writing case once more. Resting it on his knee, he wrote with handwriting that made her examine the writer to see if it was really he who wrote with such grace. He blotted the paper.

Doesn't make up medicines, she thought. Dr Newcombe would have had a bottle of stuff in his bag. This one dispenses only knowledge, written on a slip of paper. But what knowledge, if he's ignorant of the cause of fevers? I could ask . . .

The doctor stood up, offered the paper. She took it.

"Are you content there, Father? I'll send out for the medicine," she said.

"Just get me something to drink," croaked Mr Brooks. He remembered to be polite. "Thank you, doctor."

In the hall, Marian rang for the maid who produced the visitor's hat and overcoat.

"Will Dr Newcombe be back soon?" she asked.

"He's up to London until the day after tomorrow. I'll ask him to come in when he returns. If there's any change, please send for me." Leggate offered his card. "Good day to you." He took a step towards the front door. Now attentive, the maid sprang forward to open it. The doctor gave a flicker of that smile again, put on his hat, skipped down the stone steps to the pavement.

"Mr Brooks would like some tea," said Miss Brooks to the maid. "And would you take this prescription, please, to the dispensary."

Marian glanced at the doctor's card: "John Leggate, M.D., 5 Cord Lane."

Leggate . . . isn't that my aunt's new doctor? she thought. But by the docks, why ever live there? He's new, I suppose. Scientist, experiments . . . Blood letting is their principal remedy, for many of them at least, but if what he says is true? . . . can their remedies not have been subjected to scrutiny?

She went back in to her father. "Maid's bringing some tea," she said. "Would you like anything else?"

"Just the tea, then I'll try to sleep, I hope not for the last time."

"You shouldn't even think such things. The doctor said you weren't going on badly."

Her father grunted.

Always the same, she thought, imagines he's going to die. If he were to die, what freedom! . . . Wicked, wicked thing to think. She grimaced, clenching her eyes shut, chastising herself.

"Caroline will be here soon," she said to him, "but when she's gone, if you want me to read to you, let the maid know and I'll come down."

That doctor has a vigour about him, Marian thought as she went back upstairs, not just accepting conventional ideas, something mutinous about him. She smiled as this thought formed within her.

As well as her Broadwood piano, Marian's music room contained tall bookcases, a mahogany writing table, water-colours on the walls. On a small Greek column stood a bust of Aristotle. There were several chairs, a *chaise longue* upholstered in green velvet, as if inviting visitors to listen to a recital. But the room was not intended for a public: apart from painting materials at the front of the room, not fully concealed by a tapestry screen, there were books everywhere. She would look through the announcements of publishers in the endpapers of books she received, send off new orders. Her aunt, her mother's sister, freethinker, would discuss books with her, make recommendations on unsuitable subjects: libertarianism, economics.

"Miss Brooks has too many interests," Marian imagined her mother's friends discussing her. "She plays the piano, she paints, she writes essays, not as accomplishments, mind you, for that could be charming, she pursues each thing with too much determination."

"Earnestness creates a wrong impression," her mother would say. "And you dress like a governess, though when you take the trouble, you can look as presentable as anyone."

Marian picked up the book she had just finished from the middle of the carpet. *Frankenstein; or The Modern Prometheus*. People in London had read it years ago. The circumstances of its writing were now known—written after discussions of applying electricity to dead frogs to make their limbs move.

All on holiday, in Switzerland, she thought, Mary Shelley, married to the poet, they eloped, she was only eighteen, whatever must that be like?

Confined indoors, during that "ungenial" summer, she thought, great minds alive. But if you're eloping why take all those people with you, the mad Lord Byron, dangerous to know, and Mary's stepsister Claire whom she hated, why ever did she let her come? Claire fell in love with Byron, or was that later? and their doctor friend, Dr Polidori ... doctors have licence to do such things, to make experiments on animals, and the sick. Mary Shelley's idea came in a half-dream, she couldn't sleep ... it was meant to be a ghost story.

Despite her piano, and her interests, Marian seemed to herself constantly to be yearning. Should I write a ghost story? she thought. Perhaps of a doctor who makes strange experiments to discover the cause of fevers . . . unexpected effects occur, grotesque, frightening.

She leafed through the book, started to read: "None but those who have experienced them can conceive the enticements of science . . ."

Again the doorbell rang; again she ran downstairs, again overtook the maid, opened the front door to greet her friend, Caroline Struthers, a fair-haired woman, whom one might have recognized in the portrait that stood on the easel in the upper room.

She kissed her friend lightly on the cheek, took her cloak, handed it to the maid, said: "Is the kettle still on for Mr Brooks? Can you bring us up some tea too, please."

"I can't stay long," said Caroline. "Emily's got a cold, she's asleep just now, and nurse is very good, but I want to be at home when she wakes up."

"Just have a cup of tea, then I'll walk back with you," said Marian.

In Marian's music room, Caroline said: "I needed to get out, to get some fresh air. Cooped up inside, so unhealthy, and when I was sitting with Emily there was a terrible draught, right round my ankles, I hope it's not going to give me this cold, I must ask Peter to get it seen to."

"So air in great quantities is healthy—fresh air?" said Marian. "But when it's in thin currents it becomes treacherous, affects vulnerable parts like the ankles, and gets the special name of 'draught'?"

Marian looked at her friend, whose hypochondria made her always want to tease.

"You're so insensitive," said Caroline. "Has something upset you? I shan't come and visit if all you do is make fun."

"I'm sorry," said Marian, guilty now. "I'm sorry. All I mean is you're the healthiest person I know. You never get ill, but you're always worrying that you might."

"Being vigilant prevents it."

"You're right, we'll get some fresh air. We'll walk to the botanical garden, then round to your house."

"If you like."

"Forgive me?"

"I'm worried about Emily, and one can't be too careful."

"Of course . . ."

Marian, now properly penitent, reassured herself that her friend was no longer cross.

After their tea, they went downstairs. The maid, having fetched the women's winter coats, and the scarves that would protect their necks from those penetrations of cold air to which the face is immune, opened the door again. In this action she reminded Marian of the doctor, whom she had let out half an hour earlier.

The doctor's visit still agitated her, though she had not mentioned it to her friend. Relies on experiments, she thought as she followed her friend down the front steps to the pavement, calls medical remedies hogwash . . . dangerous opinions for a doctor.

Chapter Three

Nearing home after seeing Mr Brooks, Leggate thought: it's not just cholera, it's every disease, all obscure, and it's not a matter of waiting for the disease to strike and offering remedies, we must comprehend how it is communicated. "Should you not act upon it?" she says, as if a medicine would arrest the illness. They won't accept the mechanism of the body, they want us to dispense some magic stuff to abolish every ailment, pouff, like children wanting a parent to make it better. She was right though; if they realized how little we know, we doctors should all be whipped, although if ever I try to explain, they don't believe me.

Letting blood too—his thoughts continued as he turned a corner into the dock area—that's what they think doctoring's about: imagine going about opening veins with little sharp lancets and having the effrontery to charge money for it. Glad to say I don't even have a lancet in my bag. Nor any of the other fool remedies, enemas or any of that nonsense . . . what was it that Newcombe said? that one of the doctors here, Esard, Allard, what was his name? pretty much only gives enemas. Imagine, days filled with giving enemas—as if the stomach and small intestines aren't perfectly adequate for absorbing whatever needs to be administered. Glad to say I don't carry round an enema syringe either!

He opened the door of his house at the bad address near the river, and called to his housekeeper, "I'm home, Mrs Huggins." He sprang up the wooden stairs two by two, put his bag under his writing table, and lay on his couch.

She had vitality, that woman, he thought. Refreshing to find such forcefulness; what would it be like to marry such a woman, invigorating?

But in his study, thoughts and demands of patients faded. All the promptings became of cholera, and of the coming epidemic, which just last November had killed 15,000 people in Mecca.

Leggate thought again of the previous epidemic. In Sunderland, he thought, Magendie was allowed to observe, not perform post-mortem examinations. We didn't have to wait long though, two months after we're back in Paris, the disease arrives, in the middle of a carnival, with masked dancers carried moribund to the Hôtel-Dieu, and Magendie, of course, with demonstrations and pronouncements.

Leggate reflected on how, at that time, he had felt very close to Magendie, like a favourite son.

"From the moment they contract the disease they are dead though alive," Magendie would say. "Suspended! In some the pulse is weak, in most it can't be felt at all—because the heart has stopped! Therefore the vessels are void of blood. Some revive like Lazarus, but for the rest there's no awakening."

Suspended, Leggate had thought, seeking to apprehend Magendie's hypothesis. He had remembered a dormouse he had seen, brought into the zoological laboratory in winter, hibernating, its body cold, seemingly dead, but then as one watched it for a whole minute, it would take a tiny shallow breath. Is that what the disease does? he had thought: does it suspend all vital functions?

At the beginning of 1832, as soon as the disease broke out in Paris, more people lay on the floor of the Hôtel-Dieu than in beds. Magendie, tireless, patrolled the wards—devoting himself to the poorest patients, and shooing off indigent women who posed as nurses and tried to rob the incapacitated. When asked why he devoted himself in this way, Magendie replied, "The wealthy will not lack physicians."

A doctor arrived from combating the cholera in Warsaw. Ignorant of Magendie's hostility to bloodletting, he had announced that relieving the patient of blood from the temporal artery had been found to effect a cure.

Magendie, full of contempt, took a scalpel and walked to the side of a man with the pall of death upon him. The retinue followed. Magendie made sure the Polish doctor had unobstructed sight, and rested his wrist on the man's ear. He held the scalpel delicately, like a quill, just above the right temple. The man was shrunken, unconscious. Magendie glanced again towards the Polish doctor, then drew the tip of the scalpel two inches alongside the artery.

Leggate wrote in his diary:

Magendie loosened the skin around the incision, as if dissecting an animal. He passed closed forceps beneath the vessel and held it gently up so that we could see clearly that it was indeed the temporal artery. Then he took the scalpel again, and slit the vessel. Instead of a spurt of bright red, not a drop of blood fell.

When, like a miracle, no blood came from the cut artery, I could not restrain myself. The previous night, when I had the idea, I could not sleep; it was not suspended animation, it was something far simpler, and more obvious. As Magendie stood back from the dying patient to glare at the doctor from Warsaw, I said: "Could it be, sir, that the fluid from the blood has all been extruded into the gut, that the diarrhoea, which is such a prominent feature of the disease, is the leakage of fluid, the mesenteric arteries have become porous, the diarrhoea is the very fluid that is lost, so that only solid matter is left in the arteries and veins, and that's why the blood does not flow?"

I was a student, should not have spoken. Like an owl, Magendie rotated his whole head to stare at me. I saw he had not had this thought. He said: "Leggate, that is an interesting idea. You have a future. Perhaps you will be the discoverer of the cause of cholera."

Magendie believed and Leggate had come to believe, too, that no special vital force sustains us. The workings of the body are one with workings of the physical world.

"Think of your countryman, Harvey," Magendie would say. "Two hundred years ago, he showed that the heart is a pump."

Leggate was told to read Harvey's treatise *De Motu Cordis*. He read it several times, for Magendie would question him on it, to tease him, so Leggate thought. "Comparable to what happens in machines . . ." Harvey had written. It was Magendie's favourite saying.

Leggate wrote that thinking of Harvey may have been behind his own idea.

If fluid is all drained out into the alimentary canal, leaving only the solids and corpuscles in the blood vessels, no pump could work, so that is why the heart beats so feebly in this disease.

You may think these thoughts mundane: a realization that because in cholera there is diarrhoea and so little blood in the arteries and veins, the blood's fluid must have been lost in the diarrhoea, and that if the heart's beat could barely be felt this was because no pump could work properly with so viscous a fluid. You might think that every physician who saw the cholera would have had this thought. But no—this thought was Leggate's. He was conscious of conceiving it, of not having read or heard it.

Magendie made experiments on live animals, made post-mortem examinations of men and women newly dead, performed demonstrations on the almost dead. He compared humankind to the animals, created in his experiments physical effects that mimicked the causes of disease. Then he could reason that the same processes must cause the diseases themselves "comparable to what happens in machines."

In 1832 cholera was the disease for which every ambitious medical man wanted to discover the cure. But it was Leggate who came on the truth that its essential process was loss of fluid from blood vessels into the gut. Had it been widely understood at that time the idea might have been recognized as one of the twin keys to opening the double locked door that guarded the secret of cholera.

When Magendie's previous assistant had left with his degree, Leggate had been chosen from among the students to be the new assistant; he was skilled in dissection and in preparing chemical reagents. It meant he could earn a little money, but most of all it brought intimacy with the great man. Although arrogant, critical of others, given to disputes about the priority of his discoveries, Magendie was scrupulous, and although wary of his master, Leggate felt proud to be a disciple. Magendie arranged with him to perform a series of experiments to try his idea.

"If we find proof," he said, "your name will be with mine when I write up the experiments."

Magendie decided to inject mercury into the arteries of the gut in newly dead patients. If the blood vessels in the wall of the intestine had become porous, if the fluid of the blood really had leaked to cause that horrendous diarrhoea, the injected mercury would leak too and become visible as tiny silver droplets on the inner surface of the intestine. When the experiment was tried, Leggate had to inhibit his delight at seeing the minuscule beads of mercury appear. He was right!

In post-mortem experiments on other victims Magendie and Leggate injected water into the vessels of a loop of gut they had tied off. A watery substance formed, similar to the so-called "rice-water diarrhoea" so characteristic of the disease. The experiments were Magendie's. Such ideas were his genius; they swept away obscurantism—in bright flashes of discovery the workings of the body in health and disease could be glimpsed.

Impatiently Leggate awaited the writing of the paper in which his idea would be described, demonstrated in the experiment of injecting mercury, and proposed as principal cause of the cholera. His name would stand beside Magendie's.

But whereas Leggate had as yet conceived one scientific idea, Magendie had them by the dozen; always drawn to the next experiment, he was more interested in his own notions that in those of an assistant. He performed more experiments, and more yet. He was engrossed, as every scientist must be, in his own theories, so that when he came to write about cholera, he reported his experiments showing the permeability of blood vessels, although Leggate's idea of permeability as the very cause of the disease was absent. Cholera's principal cause, according to Magendie, was still the one to which he had leapt in Sunderland: cessation of the heart, "cadaverization." Leakage of the blood's fluids was secondary.

Leggate mentioned his idea once more to Magendie. He was brushed aside, and the longed-for eponym—the "Magendie–Leggate Principle"—was never coined. For Magendie, Leggate's idea was a stage, suggesting certain experiments. As in the maze at Hampton Court, it was a step along a path leading to a cul-de-sac, useful only in eliminating a possibility, of no relevance in describing the path of discovery. ·

Then occurred an incident that left in Leggate an anguished memory. He wrote an article for the leading medical journal in Paris, describing his idea—how the specific cause of the disease was that the blood vessels of the gut became permeable. He described how this explained the blue corrugation of the skin, the feebleness of the heart's pulse, the paucity of blood in the arteries . . . He hesitated about whether to mention, even by a hint, the experiments in which he had assisted his master. He decided against—it would have needed his master's permission. Leggate confined himself to his own observations, and sent off his article.

A week later, a letter arrived from the editor of the journal.

Dear M. Leggate,

Prompted by the address that you gave, at the Hôtel-Dieu, I made an enquiry, and found you to be an assistant of Professeur Magendie, a student, and also a foreigner. It is only this last fact that inclines me to believe that you may not be aware of the etiquette of sending articles to the medical press, otherwise I would even now be passing your communication to your master, rather than returning it to you. If you have speculations, observations, whatever they may be, their publication should be made by the master for whom you work, not by an assistant.

Yours etcetera

Leggate was appalled. If the editor had passed his article to Magendie, his vocation would abruptly have ceased. He could not believe he had been so foolish.

The more he thought, the more incredulous he became: he felt painfully humiliated—excruciating—and by his own action!

The inexperienced student of the great Paris hospital and of the Medical School of Collège de France, where Magendie was a king, stored his idea. He could not understand why Magendie was drawn to his inferior theory. Did he cling to it to avoid acknowledging his assistant? But in remembering his idea, Leggate remembered Magendie's words: "Perhaps you will be the discoverer of the cause of cholera."

For the disease whispered that its secrets could be known. Leggate had taken up the faith of Harvey. Magendie had laid hands on him. Leggate carried his faith: the body is a machine. As weaving machines provided clothing and dignity for the many, so Leggate reasoned, if he could discover how to tend the human machine, the body, the human condition would be improved, and he would do his part in making the world a better place.

When he had to return from Paris because his father became ill, Leggate was glad that talk of cholera was no longer on every doctor's lips. Almost as suddenly as it had come, the disease had left Europe, but so impressed upon him were its ravages in the cramped houses of Sunderland and the sordid streets of Paris, so clear was he that nothing substantial had yet been widely understood about the disease—not about its cause, not about its spread, not about its prevention, not about its cure—that he believed that when it returned, he was called to master it. For that day he had planned, and for that reason he had come to Middlethorpe.

Chapter Four

Friendship, thought Marian. That's the principle.

She was contemplating an essay for a ladies' journal, perhaps the *Illustrated Weekly Visitor*, which had taken three pieces previously, the first on this same theme.

In friendship there are distinct voices, she thought, like a fugue by Bach— a *fuga*, as the Italians say, with two voices. Neither one alone would mean so much, the inner meaning comes from their relation. Or like a novel, where in alternate chapters the reader enters first the mind of one of two principal characters, then that of the other.

Now she would extend the idea, she thought: marriage should be the perfect arrangement to realize this state, first his voice then hers, weaving into patterns that neither could accomplish alone. But what was it that nevertheless made friendship so difficult within marriage?

She sat on her *chaise longue*, wondering how she might approach the subject. She had been reading again Mary Wollstonecraft's book *A Vindication of the Rights of Woman*, to which her Aunt Mitchell had introduced her—the book which, when adolescent Christian enthusiasms had ebbed, had become for her a substitute for the scriptures. This aunt held conversation with literary friends in London. She knew Harriet Martineau, who wrote popular articles, who preached the gospel of the pessimistic Malthus that the human population of the world was multiplying itself so fast with each new generation that it would soon outstrip its supply of food.

Friendship's really the principle that makes the whole human race possible, Marian thought, the way we trust each other, the way we cooperate, otherwise nothing would work at all. Marriage is the institution that ensures reproduction, but according to Malthus we're multiplying too rapidly—so we must find in marriage something beyond mere reproduction, something more like friendship . . . And yet in marriage something works against it . . . The doorbell rang. Marian ran downstairs to let Caroline in.

Today it was again Caroline's turn to visit. Back in her music room, Marian said, "Is Emily better?"

"Much better," said Caroline. "It was just a cold. Something's going around, fortunately not too severe, or anyway not too protracted."

"My father, as well," said Marian. "Dr Newcombe came today and pronounced that 'the inflammation of the oropharyngeal region is in the phase of convalescence.'"

Caroline laughed—Newcombe was her doctor, too. "Why must you always make fun of doctors?"

"I like Newcombe," said Marian. "But their doctorish expressions—why not just say his throat's getting better?"

Marian knew that for Caroline medical pronouncement was a serious matter. Perhaps it was for this reason that she could not help teasing.

"I met a very presentable one earlier in the week, though," Marian continued. "Dr Leggate. He called to see my father when Newcombe was away."

"You didn't say."

Marian had wondered herself why she had not mentioned Leggate on the afternoon of the encounter. "I'm sorry. I wasn't sure what attitude to take."

"So you take an attitude?"

"I rather took to him. He's quite new in the town, but he's worked in Paris. He's a scientist."

"Well!" said Caroline, her voice taking on a quality of musical intonation.

In thinking again about the forms that marital relations take when they are not those of friendship, Marian's mind turned to her parents. What goes wrong? she thought. They loved each other once, so my aunt says, but how ever does one think of them as young, or in any way other than the pattern they're fixed in now, with my father so irascible?

Her mind reached back to the incident, seven or eight years ago, when the uneasiness between herself and her father became irreparable.

The uneasiness had started three or four years before that, when Marian was thirteen, when the piano began to be responsive to her. After increasing impatience Mr Brooks, who up to then had taught the piano to his daughter, had started to express overt disappointment.

In response, from devotion to her father, Marian's attitude changed to suppressed indignation.

Then had come her father's pronouncement that while Latin had been good for the rudiments of grammar, it was no longer a fit subject for a young lady. The succession of tutors who had taught her and her brother, William, at home had ceased. She had been sent to Miss Davignor's school, while William went off to boarding school in York.

There followed three years during which Marian became steadily more puritan and more pious, while she silently suffered her father's high-church belligerence from the pulpit. But she covertly continued to read omnivorously, including a good amount of entirely unsuitable matter. She continued, too, with the Latin and Greek into which the tutors had launched her.

William was glad enough of this, because in the holidays he was sent home with irksome exercises in which his sister found no difficulty. He also brought back intelligence: the provenance of the decree against feminine Latin. "In Latin there are forbidden passages, whole forbidden books," he whispered to his sister. "The Romans were lewd."

Then had come the incident: at half-term, when William was fifteen and his sister seventeen. He had brought home a copy of *The Golden Ass*, scurrilous matter tending to inflame young minds. He needed Marian still for translation.

Only a few days after the book had come into her possession, Marian was surprised that her father came into her music room unannounced. She was horrified that he saw this book open, on top of another book.

She saw him grimace. "What have you to say?"

She remained silent.

She saw him pick the book up, by a corner, wrinkling his nose at the contaminating object.

"On top of St Augustine's *Confessions*," she heard him say. "Have these two ever before been juxtaposed in such a depraved way?

"Answer me." She saw his fists clenched, veins standing out on his temples; the servants must have heard his shout, two floors below in the kitchen.

Jutting his face closer to hers, he said, "I shall confiscate every book in this room, and the piano too."

She trembled. She saw her father's disgust. Vividly to her mind came passages she had read in Apuleius's book, the lustful slave-girl Fotis... then that woman with the donkey! She had read it, unable to believe anyone could have written such a thing. But had not she herself understood it?— she was properly accused. She stared downwards, mortified, wishing the floor would open.

"You are a shameless girl, unnatural. What have you to say?"

She did not answer. Memories sprang of her father, as he used to grip the hair of her brother to shake him. Mentally she defied him to attempt this on her now; thus he would reveal himself.

"I demand you speak."

After long silence, with eyes downcast, a precept of her ideal of the previous few months, Mary Wollstonecraft, bore her up—why should female ignorance masquerade as virtuous innocence?

"It's wrong that you forbid knowledge to a human creature," she said. "I shall leave your house." She would not look him in the eye. "I shall live with my Aunt Mitchell."

Her words were almost too quiet to hear. The audacity of contemplating withdrawal to the air of her aunt's house blended with her shame's urging that her body would dissolve and seep through the floorboards.

"You will do no such thing."

Each regarded the other.

"I believe, sir, that now I am a grown woman, I must request you do not enter this room without my invitation. It is not proper, I might be ... unprepared. I believe, too, sir, that if I were to cease occupying the pew next to my mother on Sundays, if I were to live with my aunt, your parishioners would draw conclusions."

She saw the realization on her father's face, saw his lips purse as his disapproval turned now upon himself, saw apoplectic redness betray an inner battle.

"You will not defy me."

"I shall defy you no longer, sir. I shall withdraw myself."

Mr Brooks wrenched at the book that he was holding, trying to tear it in half. It was more resilient than he thought. His rheumatic fingers were not strong enough for the task. Taking the book with him, but without saying anything else, he turned and stamped along the landing, to find his wife to recruit her to his cause.

Despite Marian's piety being, by this time, much diminished by conversation with her free-thinking aunt, she prayed for forgiveness, prayed that she might love her father once more. She had been reading Augustine's confession of his fall into the sin of theatre-going and debauchery, at the same time as reading the Apuleius that her father had found.

I'm a fallen creature, she thought, reading such debased things, but without Augustine's will to rise again. I can't even respect my own father.

It was not until two hours later that her mother had come as an envoy of peace.

"Try not to provoke your father, dear," said Mrs Brooks. "I've told him it's no longer advisable to enter your room, at your age. He sees the wisdom of that . . . and my sister, you know how it distresses him that you're always in her company. I thought we had agreed."

She would have liked to repent. "I won't bring my aunt in again, Mother."

"I worry about your temperament. You're so uncompromising—as a girl, you need to be easier, more flexible, we all have to."

Marian regarded her mother's patrician bearing. "I don't see you making many compromises," she said.

"More than you think. Many more."

Then after a moment, Mrs Brooks added, "He attaches great importance to being on the right side. You shouldn't vex him by making him think you would go over to the Unitarians."

Still appalled at her own anger, and at the extent to which she felt now that she wanted nothing whatever to do with her father, Marian said, "Very well."

"Why don't we arrange for you to go abroad for a year or two. Should you like that? You've outgrown Middlethorpe. What about Switzerland? I shall make enquiries. Perhaps the year afterwards to Germany where you could study music, that way you can practise your French, and your German. Or should you like Italy, to take lessons in painting? You can take Lucy to look after you."

"He's not a bad man," her aunt had said of Mr Brooks a day after her father had discovered the scurrilous book. "He's dutiful. But when you were about thirteen, you became unwilling to do and think exactly what he wanted. You must see that men require compliance from their womenfolk. That's why he dislikes me, because I seem so incapable of it, but as his daughter you should be more accommodating."

After a day of dismay following this utterance, she had spoken again to her aunt. "With the piano I used to do everything he asked of me," she said, "but when I started to get some sense of it for myself, it was as if he couldn't bear me."

"You often see parents beside themselves when their children turn out in some way even a little different than they've imagined, and even if the difference is harmless."

Marian knew she should be more biddable—of course—and she resolved to be so, but she needed a way of tolerating her situation. It was then that it occurred to her that she would secretly, but minutely, observe the character of her father, in every detail, as if she were to give an exact account of his habits and tempers. She gave special attention to those aspects of which she could disapprove.

"I should like Marian to go to Switzerland for a year," Mrs Brooks said at dinner a week after the reprimand about the book. "She and I have discussed it. I've had a very good report of a place from Mrs Simmonds, whose daughter went there."

"Mrs Simmonds?"

Marian observed that her father knew perfectly well who Mrs Simmonds was, but employed a certain tone to be dismissive.

"Of course you remember Mr and Mrs Simmonds. From Battersby. She came into Middlethorpe today."

On the plan that her mother and she had contrived, Marian first spent a year at the finishing school in Lausanne that Celia Simmonds had attended. Then after a summer back in Middlethorpe, she went for seven months to Germany's first Conservatorium of Music in Leipzig, where she was taught by Felix Mendelssohn himself. There she heard the music of Robert Schumann, and the piano playing of Clara Schumann, his wife, and met them both. There she herself played in three Conservatorium concerts.

Leipzig was a paradise, with music on every side, with conversations enlivened by young enthusiasm. She would have stayed the whole year had it not been for a certain Andrei who played the cello with such a mixture of passion and restraint that her heart had gone out to him. Then had come the crushing incident: Andrei's betrayal. It became impossible to remain in the same room with him. Marian returned to Middlethorpe four months earlier than expected.

When Marian returned from Leipzig to her father's house, in secret mourning for Andrei, she redoubled her concentration on the piano. She decided also that she must find new material with which to sustain herself in the house to which she had consigned herself. She would overhaul her strategies for tolerating her father. It was while thumbing through Gilbert White's book, *The Natural History of Selborne*, that she thought of a new development of her idea for keeping the peace: she would study not just her father—the possibilities were exhausted—she would observe the very effects of him and her mother upon each other, in their conversations and moods. She would examine what passed between them to see how one person's character, and its response to the incidents of each day, affect and form those of the other. She would contrive small interjections herself and observe their effects too. She would make these matters the objects of observation, not just individual character but mutual effects, as her own exercise in natural history.

While Mr Brooks was all forcefulness in the pulpit, as well as to William and herself, Marian saw that he fawned on his wife, just utterly fawned, there could be no other word. Marian's mother was more worldly, more assured than her husband; daughter of a colonel, she was respected, lively in a provincial town.

Mr Brooks had become increasingly bitter since the publisher had declined his third volume of collected sermons, *Campaigns of the Christian Soldier*, and since his ambition of advancement in the Church was disappointed. By his wife, he seemed to Marian to be often humoured or, in his more disgruntled moods, sometimes merely tolerated.

Marian observed her father at one end of the dining-room table. The furniture of the room indicated the household's head clearly enough: there was a portrait of her grandfather, Mr Joshua Brooks, whose wealth, since his death, had no doubt augmented the comfort of the house in which she lived. There was his presence on the wall above the thronelike chair on which her father sat. But what went on was quite at variance with the arrangement of the furniture. It was Mrs Brooks who was the centre; it was she to whom not just servants but the family deferred.

The maid would place the tureen in front of Mrs Brooks, who would ladle the soup as one distributes largesse. "Pass this to your father."

"I believe I have made an advance, today," Mr Brooks said.

"Would you be so kind as to say grace?"

Mr Brooks did so.

"Cook has done well with the soup," said Mrs Brooks.

"I've ... em, found I think, that Napoleon made a serious mistake—but we were still not ready for him."

Mr Brooks's interest, his hobby-horse as Mrs Brooks described it to their daughter, was to reconstruct historical campaigns—those of famous military men. This he did with lead soldiers, which he had had painted specially in their proper colours, on a large baize table constructed for the purpose in an attic.

At the dinner table, Marian would see her father as she had felt when she most despised herself, wanting to win attention. She observed in him the same wideness of the eyes that she had tried to inhibit in herself as a schoolgirl, so as not to seem pleading: "Please Miss Davignor." "Please Mrs Brooks."

"Please tell me about it later," Mrs Brooks replied to her husband's appeal about Napoleon's mistake. And then: "Marian and I are planning to go with my sister to an afternoon concert in Leeds," she said. "A famous pianist,

Mr Franz Liszt, is in Yorkshire, and is giving what he calls 'recitals' of pieces of piano music in different towns. Should you like to come?"

"I don't think so," said Mr Brooks. "I'm rather busy. I think not. I've got a good deal of matter the agent's given me to look through."

Why a rector should need an agent, thought Marion, I can't imagine. He talks as if he were managing an estate.

"You'd better make up your mind," said Mrs Brooks. "He's popular, and I shall buy the tickets tomorrow. What about you, William?"

"No thank you, Mother," said her son. "Perhaps if he comes to Middlethorpe."

Mr Brooks had been a gifted organist until rheumatism and painful swelling of the finger joints had forced him to give it up. He had insisted Marian learn the piano from an early age, and had indeed taught her himself until she was thirteen, the age at which he passed instruction over to Herr Hoffman.

Mr Brooks was knowledgeable about music. Marian surmised that he would have gone to hear Liszt, but not in the company of his wife's sister whom he regarded as an enemy. William, who would have gone to a concert in the town, she knew would not relish making conversation with three women on two train journeys between Middlethorpe and Leeds.

Except for her lessons with Herr Hoffman, supplemented by occasional excursions such as the one to hear Liszt, or with her aunt to London, the years after Marian's return from Leipzig were musically meagre, for Middlethorpe is not a great musical centre.

When, at the beginning of 1846, Marian heard that Mendelssohn was to come in the autumn to England to conduct his new work—*Elijah*—at the Birmingham Festival, she thought she would write to him. Might he not come also to Middlethorpe? she thought. When I exert myself, people often respond—there's no harm in proposing it.

She wrote to Mendelssohn that she would arrange an official invitation. To her astonishment he replied that he would be pleased to come. She hurried to those who could issue the invitation. The great man graciously accepted.

The grand concert took place in the first season of the town's new theatre, the Corinthian-fronted Theatre Royal, built on the site of the former
Wilberforce Hall. Marian spoke to Caroline's uncle George Warrinder,
proprietor of *The Middlethorpe Examiner*. Articles were published, the concert
was sold out. The Philharmonic Orchestra came down from London. Herr
Mendelssohn performed his *Midsummer Night's Dream* music, and the Beethoven
Pianoforte Concerto in G, which he played to extraordinary effect.

It was just four months after this concert that Marian's fleeting encounter
with the new doctor had taken place, at her father's sickbed. Two days after
that, Marian had read in *The Middlethorpe Examiner*:

> Dr Leggate will complete the trio of judges for the
> pianoforte competition in the forthcoming Annual Fes-
> tival of Music.

The competitions, she thought. Dr Leggate, she thought. Not easy to meet
him otherwise, but perhaps the competition is a way. Science and art, he com-
bines the two. And that moral fervour when he was talking: experimenta-
tion, knowledge, principles—in her mind she rehearsed the sounds of these
sonorous ideals. Enough to fill a void?

Marian took her enquiries to Dr Newcombe.

"Leggate's father remarried after his first wife died while he was still quite
young," said Newcombe. "The new wife was French—she was Leggate's
mother, she had two sons, Leggate was the younger, but the elder died." Newcombe paused for a moment. "It's because of his mother he speaks French
as well as English."

"Did his brother's death touch him deeply?" asked Marian.

"Very much, I believe."

Marian reflected on this. "And he speaks French?"

"He attended medical school in Paris, as well as in Edinburgh," said
Newcombe.

"And are they good?"

"The leading places. He was assistant to Professor Magendie in Paris. A wonderful opportunity for a scientist."

"He's not been out much since he's been here."

"Absorbed in his research," said Newcombe. "It could be important, looking for the causes of cholera—if anyone has a chance of finding them, he has."

Marian pondered this interesting intelligence.

"I suppose, as well," Newcombe continued, "that he doesn't go out much because he's an abstainer. His father was a Methodist. But he keeps that quiet."

Newcombe smiled at Marian. When her own faith was ebbing, Marian had held an earnest conversation with him, from which she heard his reservations and his sympathy with dissenting views which he also kept quiet because the Church was tradition, however much it fell short.

"He sings at St Michael's, in the choir," said Newcombe. "He wrote an arrangement for an anthem."

"An accomplished man. So that's how he comes to judge the piano competition?"

Chapter Five

The indications can be found, Leggate said to himself, his inward voice resounding in his mind. When the next wave arrives, I shall find how the disease is communicated, and that will interlock with my idea about its mechanism—Professor Leggate! How would that be? He stifled this immodest thought.

In his study in Middlethorpe, Leggate wrote in his diary:

This is not just a disease, it is THE disease, which nullifies the very dignity of man; is it too much to suppose that in it are sequestered the secrets of our mortality, invisible but powerful, a single physiological cause? Our ignorance of it shows us still feeble, heirs to this too-vulnerable flesh.

Sometimes thoughts would be set off by an ambiguous word, which metamorphosed from one sense to the other, so after "vulnerable flesh" came this:

The flesh, Paris, scene of those events. Evette, part angel, part who knows what? Temptress—harlot many would say—but not I.

Evette was an actress. She was thirty-three when Leggate met her, while he was not quite twenty, a medical student and at that time devotee of the theatre. He saw her first in early autumn before his journey to Sunderland. He had gone to a boulevard theatre, the Ambigu, to see a melodrama. Then he had gone again, not to see the play, but to see her. They had become close, and remained so for nearly four years.

Now, towards the end of May 1847, Leggate continued to write in his diary:

In almost our last conversation, she said: "John, you are such a *naif*, you are a student, how should I marry you? What would become of us?" Was she correct? perhaps so, and this is my destiny, here in Middlethorpe? it seems that I am wading through swamps; yet what could I have done?

Suddenly she had been no longer there. She had moved from her rooms. He had searched the streets, the cafés, the theatres. She had disappeared.

When Evette left it was as if he entered spiritually the state his master described, of suspended animation. Through force of will he continued, but all impetus was lost. It was as if, in the centre of his soul, a pool had formed of some viscous fluid in which was dissolved, supersaturated, a strange chemical substance. Sometimes this fluid clarified into terrible yearning, sometimes it swirled into a vortex of betrayal, sometimes the solution crystallized out into a radiant shape, a beacon for the soul.

For months each event of each ordinary day could jolt him with new pain of remembrance. Then came a period when he thought of her only twice a day. Into the vacuum his obsession with cholera had grown, like a cancer. All else had been eradicated—the theatre, novels, architecture had all become utterly dull. His mind was filled with science only: cholera, cholera. Apart from this, and intrusive thoughts of Her, only music had remained as a soothing remnant.

Then his father became ill. He returned to London. He took over his father's office, inching forward by reading about cholera in the evenings. When his father died, Leggate wound up his father's affairs, went to Edinburgh to complete his training, came to Middlethorpe. He had reached a compromise and built an inner wall to keep off occasions for recollection. As a barrier it was not quite successful, although sometimes his inability to maintain it was no longer entirely painful.

Now I'm established, he thought, a wife isn't out of the question.

In the period of his liaison with Evette, except for his journey to Sunderland, he spent almost every night with her. Still sometimes now, as he lay on his couch in his study in Middlethorpe, a decade and a half later, despite filling his life with activity, an image would return. He remembered black hair trailed across his upturned face, soft touch, indescribable, of her upon his chest.

No, he thought, not this. I must get on.

He rose, flustered, went to his writing table, turned to stare again across the river, stirred papers ineffectually on the table, put on his coat, picked up a notebook, went downstairs.

"Mrs Huggins, I'm going out."

Chapter Six

It was when she had joined Miss Davignor's school that Marian and Caroline first became friends. Marian began to visit Caroline, saw her surrounded by brothers and sisters, nine of them, to whom she was a second mother—she was one of those fortunate souls blessed with an equable disposition, who seem perfectly at home in this world, or almost so.

But as the friendship grew, Marian started to recognize defects in her friend's family: a father given to drink, a mother almost broken. She saw it was Caroline who maintained the family's life.

"Your brothers and sisters all love you very much," said Marian. "It's not difficult to see why. You create such warmth about you."

"As the eldest of a bunch of ragamuffins, it's the only solution," said Caroline. "One acts in self-preservation."

"You bear it bravely."

"There's no alternative . . . but it helps to talk a little . . . about some of the difficulties . . . to you." Marian wondered if her friend were about to cry, and reached out to touch her wrist.

Marian and Caroline began to confide to each other what was closest to them. Marian found herself reflecting on her own family, articulating her exasperation with her father: "It is not as bad for me, I know," she said. "My father is respected, but disapproving, always disapproving . . . at least your father shows you some affection."

"At the same time ignoring my mother, but I suppose he does his best."

"I have flashes of anger," said Marian. "More than flashes—yesterday I spent all day feeling so furious towards my father I couldn't speak to him."

Caroline's thoughts flowed always from her amiable temper. "This book by Mary Wollstonecraft that you lent me," she said—she was giving it back to

Marian—"Don't you find her harsh? 'Vituperation on the rights of woman,' I'd call it."

"You don't find an echo in yourself, of what she says, or find it stirring?" asked Marian.

"In places I do, and I expect she's right, but so theoretical," said Caroline.

"I suppose I'm rather theoretical myself."

"This bit here, for instance, '... strength of body and mind are sacrificed to libertine notions of beauty, to the desire of establishing themselves—the only way women can rise in the world—by marriage.'"

"Well?"

"If I look presentable in a new dress, why should I be called 'libertine'? And I shall not think badly of being married."

"Despite what you see in your own family."

Caroline looked shocked.

"I'm sorry, I'm sorry," said Marian. "I didn't mean that, I meant your mother, by your own account, has not much companionship in her marriage."

"But surely," said Caroline, "it is possible for married people to love each other. If men and women are different, we women have the best of it. I wouldn't want to be scribbling through yards of ledgers in a bank, or off at sea for months on end. Of course marriage sometimes turns out ... not as well as it should, but that's for different reasons."

Each knew she was attracted to the other by something admired, but felt to be lacking in herself. Marian was moved by Caroline's feminine cheerfulness, her easy acceptance of others despite their faults, despite difficulties in her own life. She thought Caroline liked her because she refused to accept constraints too easily. Both remained silent after what Caroline had last said.

Marian's inner world was actuated by a constant wonderment and searching, now thinking about this, now earnestly pursuing that. At this moment, prompted by her friend's faith in love, Marian wondered about her own character and about that of her friend, wondered how one might find a point of balance between such irreconcilables, between the amenable and the recalcitrant.

When, after Marian's return from Leipzig, she and Caroline attended dances, they acted in unison. Accompanied by their mothers, they would sit together. On the first occasion, as the little orchestra began, they were approached by a young man they knew slightly. "Good evening," he said, looking towards the fair-haired one. "Miss Marsh, may I have the pleasure?"

"I thank you for your kindness. For now Miss Brooks and I will observe." They had agreed to dance only if approached by two young men. Marian had been relieved by this arrangement. Her tendency was to dress plainly, though on this occasion her mother helped her choose something elegant. Even so, exhibited like this, she felt like the poor relation beside her friend.

The young man went to join a knot of others at the opposite side of the room.

"Whispering and peering at us," said Caroline. "They've no idea how to behave."

"I'm not sure I've any more idea than they," said Marian. "Here we all are. We've all known each other for most of our lives, them and us, and now we all have to behave in an utterly different way."

"In a small town, of course one knows everybody," said Caroline, "but that's no reason for them not to behave properly."

Marian and Caroline regarded the room; they spoke in clear tones eschewing any movement that might provoke schoolgirl giggling. Another young man whom they knew approached. He was sent off like the first.

A little later Edward Warrinder came towards them: a large man, who reminded Marian of a bear—he was Caroline's cousin, known also to Marian's family. He was accompanied by a friend. "We wanted to ask if you would honour us," he said.

"We would take it as a great favour," said the friend.

The young ladies looked at each other. "We should be pleased," said Caroline. The pattern of Marian's and Caroline's appearances was established. Young men learned that two partners were needed.

When, in the second year of these rituals, Warrinder approached Marian and Caroline, accompanied by Peter Struthers, of good family and comfortable income, and when Caroline started to encourage Mr Struthers's courtship, Marian tried not to mind that her friend's new concerns dominated her conversation.

"I should live near my family," said Caroline. "And there's a house that would be quite suitable, would be ideal, in fact. And now that Frank is grown, and Libby is eighteen, she can take my place at home, she's more sensible than me in many ways, and more capable, though Hetty will always be a trial for her, but I would still be able to see them all every day . . ."

"But you would soon start your own family."

"The little ones would like that, like having a new brother or sister, and I'll be just round the corner."

To compensate, Marian found herself giving more attention to Mr Warrinder.

"May I compliment you?" he might say to her. "You appear this evening as the very embodiment of feminine radiance."

"You are flirting with me, Mr Warrinder," Marian might reply.

"I am wounded that you don't accept my compliment."

"I accept it on behalf of my dressmaker, and my maid, Lucy, who pinned up my hair."

"I speak of the picture not the frame."

"Imagine I wished to compliment you, sir," she said. "If the occasion were to arise I should like to speak properly. Should I comment upon the manly whiteness of your shirt front?"

"You could compliment me on my good taste in preferring to converse with you rather than anyone else in the room."

"Let me see if I can guess what comes next. Now you'll suggest that we dance."

"I shall."

Occasions for meeting become more frequent. Marian found that Warrinder had fallen into an enjoyable kind of raillery with her, at least that is how it seemed.

"Would you care for a wager as to when Struthers and Caroline become engaged?" she heard him saying. "Before the end of the month, I'd say."

"I distrust such offers. The proposer invariably has covert knowledge."

"No—I assure you—I merely observe."

"I have the advantage of direct communication with one of the princi-pals," said Marian. "I must advise you against losing your money."

"There is no one to whom I would more willingly part with it."

"Now you wish to buy my interest in your conversation?"

"I think Struthers and Caroline will do well. He is ambitious. She'll make the perfect wife for him. A man needs the complement of a wife and the comfort of home. She makes his endeavours worthwhile."

"When school finishes, a woman is the prize."

"I mean that a man in himself is very incomplete. Aren't you interested to know how incomplete we are?"

"I am fascinated to learn that our part in relation to the opposite sex should be to repair a defect."

Then, gathering her courage, Marian said, "I have just had an article pub-lished that deals with relations between the sexes."

"Indeed," he said.

Marian thought he looked impressed. "It's not in a publication you would read," she said. "But if you call round, I'll leave out a copy of the journal for you, and let the maid know that you'll pick it up. If you want to, that is."

She was pleased with her first publication, pleased also with the oppor-tunity to tell Warrinder about it. Apart from her family and Caroline, she had not been sure who else she could tell directly without seeming to boast or to be making some sort of point. Writing, it seemed, was an activity in which it was polite to wait until other people noticed something one had accomplished.

Marian found from the maid that the journal had been picked up, and two days later, Warrinder called at the rectory, and asked Mrs Brooks whether he might speak to her daughter, alone. Marian received him.

"I've been thinking a good deal, and I've come for a particular purpose," he said.

"A purpose," said Marian, with a sense of foreboding.

"I read your article," he said. "I found it very meaningful."

"You did?"

"On the subject of friendship between the sexes. I should like us to be especially close friends."

Marian's foreboding burst into certainty. He's going to propose, she thought. Somehow he took my article as an indication that I'd be willing to accept . . . All these weeks, all that banter, why didn't I think he might take it like this?

"I'd very much like to hear what you thought of my article, its argument I mean," she said.

Marian saw Warrinder hesitate. He's unsure of how the moment is to be managed, she thought. Thank goodness he allows himself to be diverted, if only briefly.

"I congratulate you on it," he said. "It's uncommonly well expressed . . . I've thought I might take up writing myself, and if I were to do so I'd value your views, constructive of course, and help. I'd very much value views from someone of such discernment . . . from you, I mean."

"But what did you make of my argument," she asked, "that in friendship between the sexes there should be the most complete equality, as between two independent beings?"

"An equality of expressing views, and of offering replies, I've always thought so."

"Then I'm pleased, though when you said you might take up writing, and that you would value my discernment, I became worried: I imagined you thinking that you might write and I might be a person who offered help."

"Would that be so wrong?"

"I wondered why you didn't imagine that I would write, and you might help."

"Of course . . . that too, of course. But I meant I might write seriously, for a living, not just in an occasional way as with your article."

"I see."

"But Miss Brooks . . . Marian."

She saw he was bringing himself to the point.

"You must know why I'm here," he said. "I share the views of which you wrote, about friendship in marriage. I took the fact that you gave me your article as a sign—it would make my life complete, for you to be my very closest friend, my wife."

Marian stiffened as he took a step towards her, saw him holding out the copy of *The Illustrated Monthly Visitor*, to return it. If she were to take it, both his hands would be free. She saw how they would then move to her upper arms, saw that his face, bending over hers, would be inches away.

She did not take the magazine. She moved back a pace, sat in an armchair.

"Please," she said, "please sit down, we needn't stand. I feel quite taken aback, I scarcely know what to say."

He remained standing.

"I am honoured . . ." she said. She glanced up, conscious of him, large before her. "But I wonder whether you do understand," she said. "I couldn't be honest in entering any contract to obey, as it says in the marriage service. Did you read my argument about that? It is the centre of the article."

"You were writing to provoke," he said. "Your women readers must have found it witty—'in cases where there is a correct answer—the proper time for dinner for instance—wives should take instruction from their husbands, but where it is permissible to express a preference or an opinion then wives should certainly be allowed to do so.' That must have raised a smile."

"But the conclusion, whether in the end the man should be the master?"

"You were teasing, you didn't mean to cast doubt on that, surely . . ."

She saw him searching her face for the smile that would confirm the sense of humour that he believed he recognized in her article. I shall keep a straight face, she thought.

"You're making fun," he continued. "Surely . . . it's the very basis of society, anything else would be unnatural, you don't question that men have to decide what's best in the world, where we must act, and provide . . . do you? It's universal, surely, that the woman rules in the home—arrangements for dinner . . . and over children and servants . . ."

"My article miscarried," she said. "You see how clear it is! It's the woman who needs help in composition. You show that what I wrote is so open to misunderstanding as to convey the opposite of what I intended."

They were silent.

"You think I'm a shrewish Kate, asking to be tamed," she said. "But I contest the customary conditions of marriage. That's what I'm saying in the article . . . not any specific offer, but . . . I am sorry, you must forgive me, I thank you for your kindness, I'm not cut out for marriage, not in general, I mean. I'm very sorry."

She rose hurriedly, eyes downcast, flinched away from the pressure of a restraining hand on her arm.

She went rapidly to her music room, closed the door, leaned against it from the inside.

I'd be overborne, she thought. He likes to dominate.

She closed her eyes, remembered a waltz, the bulk of his chest, when he had tried to press her too close for politeness. She put her hand to her throat, as if to shield herself.

Forceful . . . but to be mastered by him? she thought.

She pressed her lips together. She was reminded of the pursed lips of her father—as if habits of gesture could be passed from parent to child. She changed her expression, went to her writing desk to renew her determination on paper. She ended her letter like this:

Please forgive my contrariness. Consider me a sport of nature, too spiky for a wife for you or anyone, but if you now respect me, as your declaration implies, please look favourably on me, although I presume on your good nature. Please continue our friendship, because friendship is the ideal most close to my heart.

She signed the letter, and sat pensive. Am I honest? she thought. Should I say that his bigness frightens me, that I could not bear . . . those aspects of marriage with him.

An image came to her, unbidden, of Andrei, the cellist in Leipzig. He was a fellow soul, she thought. Or so I believed . . . friendship, shared love of music . . . and I thought he loved me, that he felt the same, then he becomes engaged to that Grete.

Marian's face flushed with the memory. Her heart had gone out to him, while he, quite, quite oblivious, apparently not having considered her for an instant in that way, had seemed to find nothing incongruous in allying himself with his landlady's daughter.

"Ough," she said out loud to herself. Now this with Warrinder, she thought.

For the most part she tried to avoid thinking of the humiliating scene of Andrei, all smiles.

"I would like you to meet my very good friend Marian," he had said to Grete.

"She agrees to become my wife," he had said to Marian with his siren Russian accent.

"Ough." Marian made the sound again.

Standing beside his Dresden doll, she thought. And the doll herself, yellow-haired, plump-bosomed, complacent . . . then he expects me to be polite!

And now here I am once more, but this time it is me to whom the proposal is made, a real proposal. But I can't say Yes. The word that forces itself on me is No. Marriage is not the relation I want . . . not with Warrinder.

I'm not being honest. It's not just the general case. There's something in me that doesn't consent to him, in himself. I do value his friendship. But how could I say there's something about him I reject?

Warrinder replied to her letter, saying that he was much dejected by her response, that he hoped that she might reconsider, but he said that he regarded it as a courageous gesture for a woman to stand out against marriage.

Within the next months, Marian noticed she had not been unique, nor Warrinder so cast down as to eschew other female companionship. After a while, after several awkward meetings, the breach between them was, in a manner, repaired. Marian seemed to have succeeded in her aim, for she and Warrinder became able to converse as freely as before, although for her now without fear of that prickling sense of an expected obligation that had so suddenly appeared between them.

The arrangement had other effects on her. It was as if she had stood on an upturned barrel in George Square, and declared herself for spinsterhood.

Chapter Seven

In the unsalubrious neighbourhood of Crusoe Street, Leggate stood on the pavement, writing in his notebook.

"Dr Leggate, Dr Leggate . . ." A young woman, well dressed, came running behind him. "I saw you from the other end of the street." She was out of breath. "I have been looking all over . . . Millie has too. Someone told me they'd seen you. Can you come? It's Connie, her waters have broken."

"Isn't there a midwife?"

"She refuses to attend, sir. She . . ."

"Very well, I'll come."

The woman who had called to him was Amy Gold, a fair-haired woman with an elfin face. The one to whom he was summoned was Connie Smith. She and Amy were his patients. Three weeks after arriving in the town he had returned home to interrupt the housekeeper whom he had inherited from his predecessor. "The doctor can't see your mother," the housekeeper was saying to Miss Gold. "He's on his calls."

"Then I'll wait."

"Wait all you want, he still won't see you."

"My money's as good as other people's."

"I'm back now," Leggate interrupted. "You can speak to me. Thank you, Mrs Thwaite."

The housekeeper scowled, retreated to her kitchen. When he came back from seeing Mrs Gold, she accosted him. "Don't you know what kind of woman that is?"

"What kind of woman is it?" he asked mildly.

"You may be an educated man, doctor. All I can say is that if you don't know that, you don't know much. She's a hussy, she's a . . ."

"Are you saying I shouldn't attend her mother?"

"I am telling you that if those is the patients you take on, you can forget respectable folks. We may be by the docks, but we don't have to sink that low."

"Really, Mrs Thwaite, I think as a doctor I shall decide who needs my services." Mrs Thwaite was replaced by Mrs Huggins. Leggate continued to attend Mrs Gold, and later her daughter Amy, and Amy's friends.

Halfway along Friars Walk, Leggate and Amy came to a house which they entered. To police and to reformers in the town it was a brothel. We may have been given by public reputation or literature an idea of a brothel as containing a luxurious room in which young women, sparsely clad, lounge in postures of languid invitation. If so we should have been surprised. The house was more like a common lodging house, although less sordid than most. The rooms, except for the kitchen and a room in which lived the housekeeper with her ten-year-old son who carried coal and did errands, were each occupied by just one person. These were not women who stood at street corners, nevertheless they lived by keeping company with men.

The kitchen was on the ground floor, opening directly from the hall. It was used in common by the six women who lived there. On one side was an iron stove, in the middle a large table of scrubbed pine planks with gate-back chairs around. The room was not uncomfortable, although not used by gentlemen visitors. Leggate did not think it incongruous that a jar with fresh forget-me-nots stood near a teapot on the table. Amy took Leggate up a flight of stairs, to the room of Connie, the woman in labour. He spoke to her, while Amy went down to the kitchen for a jug of water so that he could wash his hands. He made an examination.

"Will this be your first baby?"

"It will, sir, if I make it."

"It won't be born for two or three hours at the soonest," said Leggate. "I'll fetch my things, and be back directly. I'll bring something for the pains."

She's in mortal terror, he thought. She's right, one in twenty do die after parturition . . . but even if nothing happens for several hours, my presence will reassure her.

Any who saw Leggate setting out from the women's lodging towards his own house would have laughed. Gentlemen generally walk with an evident dignity—now here was such a one, well above thirty, trotting briskly through the streets, like nothing so much as an eager young pony.

In half an hour, Leggate was back at the lodging house, where he spent the rest of the day and the evening. The labour was not unusually long, but Connie was slight, her pains intense and harrowing for all. It was as if they were embarked on a stormy voyage of unknown duration, to a destination they dared not guess. Amy was horrified. She had never previously been present during a labour.

A new life entered the world: a baby girl. At three in the morning Leggate returned home.

Chapter Eight

At the beginning of May 1847, Marian read this in *The Middlethorpe Examiner*:

HORRIBLE SLAYING DESECRATES
CHURCH AND SABBATH

On Sunday last, persons early in the streets, including those on their way to St Michael and All Angels, were confronted by a fearsome sight. A young woman, with her delicate lavender costume in perfect array, was found with arms akimbo, and with a dagger thrust into her heart, on the flagstones at the west door of our principal church. A member of the gathering crowd was delegated to summon a constable. By the time the constable's official presence became known at the scene, before an inspection had been made of the body, and after he had departed, and come again with some of his fellows, accompanied by an undertaker who would transport the mortal remains of the unfortunate woman, the corpse must have been seen by nigh on two hundred people . . .

The article was written by Edward Warrinder. After some casting about for an occupation, and evidently acting on the notion of a career in writing that three years earlier he had mentioned to Marian, he had applied to his uncle, *The Examiner*'s proprietor, to work for the newspaper. His first job was to gather intelligence of shipping in the port. He progressed to writing articles of general news. Then came the opportunity of the Sunday murder. Though some said the article's tone was too close to irony, for most people the article carried a mixture of titillation and irreverence they found inviting.

The principal policeman in charge of the newly established department of detection at the Middlethorpe Police Station was a friend of Warrinder, and had given him a description of the young woman to print in the newspaper, along with a request that anyone who knew her identity should communicate with the police station.

The cogs of rumour turned: a murderer stalked the town. Ladies were terrified to go out of doors. Young gentlemen declared intentions to hunt down the assassin of this unknown woman.

Edward Warrinder lost no time in letting everyone know that he was the author of the articles about the murder.

"Your cousin is making a name," said Marian.

"Some people think it tasteless," said Caroline apologetically, as if wondering whether Marian's interest in Mr Warrinder were sparked once more.

"He's found a stage for himself. We'll see if he plays Mr Punch or the policeman."

Two weeks later Marian sat with her mother at embroidery. She did not much like embroidery, peremptory piano practice allowed her to avoid it for the most part, but it permitted an enjoyably sporadic kind of conversation.

"I think I'll enter the piano competition again this year," she said. She had not played in the town competitions since her return from Leipzig.

"That would be lovely," said Mrs Brooks. "I liked it when you played in the competitions. Isn't the closing date for entries soon? I remember seeing it."

"At the end of this week, Mamma, and the competition's next month, the 21st of June."

The two women were silent for two minutes. "Haven't you left it late to prepare?"

"I hadn't decided until now whether I should enter . . . or not definitely."

Her mother glanced at her questioningly. Marian saw her wondering what the reason might be that she had not mentioned the competition until now. Marian had avoided telling her mother about her intention until the last minute; it was—she thought—because she despised lying. She had wanted to postpone until the last moment answering enquiries from a perceptive mother about her reasons for entering the town's competitions again.

William came into the room. He carried an issue of *The Middlethorpe Examiner*. He moved an armchair to the window, then walked between the two women to retrieve a footstool, returned between them again, sank into the chair, his boots on the footstool.

"I hope your boots are clean, William," his mother said.

"I am sure, Mother, that whatever slight mark my boot might make will be quickly removed by the parlour-maid."

Mrs Brooks seemed resigned. "I had a letter from Uncle James, in Canada. He says you have a new cousin," she said.

Mrs Brooks's brother James had gone abroad when young. He had married and set up in some substantial way in Upper Canada.

"Whenever we get news, it is of some accomplishment," said Marian.

"I don't think I could be doing with the Colonies," said Mrs Brooks. "What will you play, dear, for the competition?"

"That woman found dead in front of St Michael's—a woman of ill repute . . ." William attempted a change of subject by offering this phrase from *The Examiner*. He had found something to shock his mother and sister. It had shocked the whole town. "Is this what Middlethorpe has come to?" he said. "Publishing a scandal sheet? And our friend Warrinder's behind it I hear."

"Warrinder's not our friend," said Marian. "If anything he's my friend. You don't much like him."

"And shall you continue to like him, in this new career?"

"That's last week's paper," replied Marian. "You're a week out of date."

"Ah," said William. "You mean the assassin has been flung into the town dungeon, where even now he languishes?"

"Here's this week's news," said Marian, reaching for the paper beside her chair. William came to take it, started to read.

"In the competition I shall play a Chopin piece not heard before in the town," said Marian.

"That French pianist whom you've been playing so much recently?" said her mother.

"He's Polish." William spoke again, from behind his fresh newspaper. "He'd like to compose for orchestras, but doesn't quite know how to get all the instruments to play together. He shows off to royalty, and has love affairs with women who pretend to be men."

"You are a perfect Philistine," said Marian. "Did you leave Oxford because they had no more to teach you about talking on subjects of which you know nothing?"

But William now had his mother's attention. "Whatever do you mean?" she asked. "Has love affairs with women who pretend to be men?"

"M. Chopin had a liaison with a famous author," Marian said, "a woman who writes under the name of George Sand. William's sneers would have more cogency if he did something other than laze about with newspapers in the morning, ogle at horses in the afternoon, and lose money at billiards in the evening."

Attention shifted. "Have you been losing money again, William?"

"No, Mother, I have not."

Now Marian asked, "Would you do your sister the kindness of collecting the papers for the piano competition?"

"You could manage that yourself. A maid can walk with you," replied William.

"Marian doesn't want to run about the town," said their mother. "You'll be going out. It'll be no trouble for you."

William seemed about to reply, but decided against. Instead he offered to walk with his sister.

The atmosphere between them when they were alone together was quite different from their exchanges in the presence of their mother or father.

"So you are going in for the competitions again?" he said. "You'll show up all the rest."

"It gives me something to aim for. You go off to Oxford—you could go to the Antipodes. You have no idea how restricting it is for a girl. Music is some commerce with the world."

"Oxford's not so much. Latin's rightly called a dead language."

"You say that to be spiteful. You know I'd love to study Latin with learned people. You should finish, in any case."

"It would be hypocritical for me to be a curate. I'd be content with a small farm—perhaps a farm of moderate size—I could raise horses."

"Father will buy a farm for you. Your wish will be granted."

"He won't buy a farm unless he thinks the terms advantageous. My wishes will be unimportant. I'm nothing but a disappointment—you're the clever one, you've always been the favourite."

Marian looked at her brother. "Whatever's in your head all the time? I wonder you can't see it. When we were young, I think he did like me, but he doesn't really like women, except Mother. Since I started to grow up, he's been nothing but disapproval towards me. He disapproves of my playing romantic French music, of my reading books that Aunt Mitchell recommends, of my writing pieces for silly ladies' journals. There's no real hope for me, but he's not given up hope in you. You'll do something respectable that'll satisfy him, even though it won't be the church or the army, you'll own some land, and be someone."

William glanced at his elder sister. He said, "I don't know . . ."

"Here," said Marian. "This is the place. I'll pick up the papers. We can walk along St Paul's Street, and then beside the river for a little. Would you mind that?"

In the issue of *The Middlethorpe Examiner* that William had first started to read when he had joined his mother and sister, the identity had been announced of the murdered woman, Marie Watson. In that issue, which appeared a week after the one in which the case was first published, she was described as the illegitimate daughter of a mulatto mother and a plantation owner in the West Indies. At the inquest a word was used to describe Marie Watson's profession that is seen in print only in reports of the police courts. Edward Warrinder wrote more extensively than was necessary about the life of such women. To Middlethorpe's newspaper-reading classes, he brought news from a world beyond the boundary of society.

Marian's father had taken a dislike to Warrinder when he was courting Marian. He had an agreement, so it seemed, that his daughter would marry the son of his friend Colonel Willoughby, so although the courtship of young Willoughby had been neither energetic, nor accompanied by palpable sign of success, Mr Brooks was in no doubt about opposing any alternative scheme. Now he was glad of the excuse of newspaper articles to tell his family to cease all familiarity with Warrinder.

Mrs Brooks saw no reason to comply. She outraged her husband by conversing with Warrinder after morning service.

At a dance at the beginning of June, Warrinder insisted on partnering Marian on three occasions.

He's clever enough, she thought, and he has gumption, but imagine if I had accepted! The price is that I'm an old maid. He continues to be my friend, so long as I don't look at anyone else.

Marian shuddered—at the predicament she had escaped of being wife to one whose newspaper articles seemed calculated to provoke salacious gossip, and at the alternative predicament of loneliness. Now that the case of Marie Watson was on everyone's lips, she found herself thinking of Warrinder perhaps more than previously.

Chapter Nine

Soon after he had arrived in the town, Leggate had been invited to dinner by Newcombe. There he met the Halliwells—an affinity between him and Thomas Halliwell immediately affected both. Leggate found himself accepted into their family.

Halliwell was owner with his brother Robert of a factory for making pumps, Halliwell Brothers. They made pumps for ships, for wells, and for windmills. They supervised drainage and the supply of water for irrigation and for manufacture.

"We're a new profession," Halliwell would say, "engineers, who know about machines."

Halliwell was untidy, careless in dress and manner—but incessantly busy. Leggate was always surprised his friend ever found time to chat.

"What have you been up to?" said Halliwell.

"More maps," replied Leggate. "With the pattern of the disease in 1832, I'll see whether it's the same or different when it returns. That's my trap."

"Perhaps it slips through traps," said Halliwell. "It may not depend on local conditions."

"It seems to be affected by places, but it also comes with trade. With trade, there comes this evil."

"Perhaps God is telling us something about the ways of trade," Halliwell laughed.

To Leggate this was nothing to joke about. "No, it's not like that; it is a duty to find out, because it affects mainly the poor. We can scarcely hold them to blame."

"I'm sorry. I don't mean to make light."

"I know . . ."

"And I can't bear to think of my children in danger . . . I've got something to show you. Come."

Leggate followed his friend into a workshop built onto the back of the house. It was filled with woods and metals of different kinds, with hand tools, and a lathe worked by a treadle. Had you thought that Halliwell's untidy person would imply a chaotic place of work, you would have been wrong. The place, though utilitarian, was exquisite in its orderliness.

"Look at this," he said. He held up a small, shiny device, three inches long. Its main parts were a brass cylinder and a spoked wheel with a thick, round rim. A shiny tube, about three-sixteenths of an inch in diameter, had been brazed into one end of the cylinder. Halliwell blew into the tube—noiselessly the flywheel started whizzing, propelled by a little piston rod.

"My latest model. If I scale it up twenty times it's a steam engine to power my pumps. Now here's the real beauty of it."

Halliwell beckoned Leggate to follow him. They left the workshop, and went to the kitchen.

"Excuse me, Mrs Perkins," Halliwell said to the cook. "I just need some water. May I take this?"

Without waiting for a reply he poured water from a jug into a bowl, then immersed the end of his machine's little tube and rotated the flywheel rapidly with his finger.

"Now it's a pump," he said, as a string of bubbles rose through the water. "The principle's the same. With an intake tube, I could make larger ones to pump water. They both work with a piston. It can either create motion to do work, or the circular motion of the wheel creates a pumping action. It's a matter of getting the valves right."

Halliwell's smile spread across his face, and little creases marked his upper cheeks from the habit of this movement. He put the bowl down, but carelessly, so that it was not fully on the table. As he turned, he brushed it with his jacket, so that the bowl fell on the flagstone floor to start a rivulet of water among scattered shards.

His face changed to concern. "I'm terribly sorry, Mrs Perkins, I've dropped your bowl, here let me . . ."

Mrs Perkins interrupted his retrieval of the broken pieces. "Don't fret, Mr Halliwell, I'll see to it."

Halliwell took his friend off, his train of thought uninterrupted, still full of talk.

"I went to see the new hydraulic tower and the waterworks," he said. "They had the ceremonial opening yesterday. Mr Taylor was there, the man from London who drew up the scheme and had the force pumps brought in from Birmingham. I couldn't have done better. They are huge and impressive, though the scheme as a whole is a mistake."

"It's all working?"

"As far as it will ever work. I talked to the surveyor, Mr McPherson, not a bad chap. The reservoir holds five million gallons, enough for a twenty-four-hour supply for the whole town. There it settles, and passes through filters, then they pump it up into the tower. There's only a four-foot fall per mile from the river there to the estuary, but a 160-foot head of water in the tower."

"The reservoir gets filled only with fresh water, not salty?"

"That's the idea. They open the sluice from the river into the reservoir for two hours at the bottom of each ebb tide. They say they avoid collecting brackish water. They'd fired up the engine, then the climax was when they started up both pumps."

"So that's the water which comes into town now?"

"The main twenty-two-inch pipe was laid a while ago, and most of the pipes for houses and standpipes are in. The town's in four sections, each section to be supplied for three hours a day."

"I've had a cistern put in, and the pipe's installed," said Leggate.

"I don't like the idea of everyone needing a cistern, it's far from ideal, but the scheme works as designed. What more can an engineer say?

"The mayor announced: 'This stupendous advance, providing water for the town's entire needs for the foreseeable future, ending the haphazardness of our supplies.'

"Mr Marvell, whom I saw there, whispered that he doubted the filters would have any effect. 'If you filter pea soup, you still have pea soup,' he said."

Halliwell's eyes met Leggate's seeking the reflection of his sentiment.

"I don't like it," he continued. "The water for our house here comes from Tafts Spring. I've refused the new supply. I'm sure the spring water has beneficial qualities. When I mentioned this to McPherson, he said I was being superstitious."

He broke off. "I want to show this to Sophie," he said. With his little steam-engine pump, he headed for the drawing room. His eldest daughter, Sophie, sat with his doe-eyed wife, as they each constructed an intricate pattern of embroidery within a circular frame.

"I said I'd make you something to show how a steam engine works," he said to his daughter. "You be the steam by blowing into this tube."

He handed the device to Sophie, who started to blow into the little tube. Nothing happened.

"Hold it so that the flywheel can go round." She blew again. The little wheel whirred.

She giggled, alert and excited: "This is what those machines are like?"

"Just like that. Tomorrow I shall show you how it's made."

"So this is for me?"

"For you. First of a new line."

By half past nine Sophie was in bed.

"Thomas says your researches are going well," said Margaret, a gentle soul whose silky wisps of pale brown hair were always too soft to stay in place.

Leggate looked at her. A little gust of laughter crossed his face: "He means I set off in three directions at once, and then get lost."

"Did you see someone calling himself 'Common Sense' wrote to the paper?" Margaret said—despite her mildness, she liked nudging Leggate into talk. "I forgot, you hardly read the newspaper except to find out about the cholera. Anyway this is about the cholera. Mr Common Sense wrote that when ships arrive from any infected district, every suspicious package would have to be opened to see if it contained clothes worn by anyone who died, and then they should be destroyed."

"Hurrah for contagion," said Leggate. "Obviously a doctor, he already knows the answer. It's called medical opinion, price half a guinea. They're all so certain, they can't think it needs to be turned into a question."

"I don't see..."

"Or two questions: if it's contagion spread by touching the victim or the victim's clothes, how many cases are explained by this? But even more important, how many are not?"

"The newspaper seems certain the cholera is coming."

"That part is right."

The three were silent, each thinking of what a new epidemic might mean.

"In the paper too, the poor woman murdered," Margaret diverted herself from the distressing thought she had started in herself. "You know some of those women, John?"

"One of her friends told me an inquest had been held, so I read the report in the paper. The man who did the post-mortem is a charlatan."

"Dr Tuff," said Halliwell.

"He said the woman was killed by strychnine," said Leggate. "If no blood was spattered by the dagger then that didn't kill her, so he got that much right, but I can't imagine how he thinks she was poisoned by strychnine."

"I thought he found some traces," said Margaret. "Isn't that what he said?"

"Strychnine produces convulsions, so her limbs would be all contracted, not stretched out as she was found. And even if it had been strychnine, he couldn't possibly know it from poking about in the intestines."

Halliwell looked affectionately at his wife, usually reticent, now animated.

"The woman herself," she said. "You've got some of these women in your practice?"

Leggate hesitated... "I'm not a prude," Margaret said. "The paper said she was a prostitute. Surely not the only one in town."

"She was my patient, she and the other women in that house—it's ridiculous for the paper to make so much of it, people will make it last for months, they adore gossip."

"Like we're doing now, in talking about it!"

"I mean the hullabaloo about the killing of one person, the idea of whom titillates, when dozens in the town die every day of poverty. People aren't interested in her, or the troubles of her life; what they love is the scandal."

"Do you think it's sensible to have such women in your practice?"

"Prostitute's the term used to denigrate any woman who's unmarried and lives neither in the workhouse nor with her parents."

"But there are some..."

"Of course there are, and sometimes, like anyone else, they need a doctor. I do more good there than with many of my patients." He laughed to prevent his tone becoming self-righteous. He mimicked a matron of the better class: "Couldn't you come and rub in some of that soothing embrocation which that very nice Dr Driscoll is prescribing."

Halliwell laughed.

"But I'm sure you're right that it must be ruinous for my practice," he said.

Margaret was looking so unhappy that he said, "You think I shouldn't be flippant."

"I worry there's something wilful in it, that you'll do yourself some harm."

"I find it admirable that while others think like sheep, he thinks for himself," said Halliwell. "We're all very fortunate—why should the unfortunate be deprived of doctoring, as well as of everything else?"

"You both know what I mean," said Margaret.

Her husband continued, "And always questioning to see what's beneath the surface of things. I read about the inquest. It didn't occur to me there was anything wrong."

Chapter Ten

When Marian had received, in October of 1846, the edition of Chopin's works from Gebethner and Wolff, the technical difficulty of many of the pieces had awed her.

These are for his own performances, she thought, or only for the most accomplished musicians of Paris or Leipzig. Can I do this?

For several days she tried out themes, tested left-hand chords, seeing what her ability would allow. Her fingers were long, thumbs jointed in such a way that she could play tenths without difficulty.

Such invention, she thought. Composing's another realm than playing, but he does both, writes as well as plays. They all do, Mendelssohn, Liszt, Clara Schumann, all of them.

She took the nocturnes and études to show Herr Hoffman. He had schooled her on Bach and Beethoven—would he say something dismissive about these pieces? When she gave them to him, he turned the pages slowly, stared at the staves for fully five minutes. Marian waited, not even a clearing of her throat able to impinge on his concentration.

"For a week, please, I may keep these?"

Herr Hoffman said he would prepare two pieces so that she could hear them played. He would mark, too, for Marian the fingerings, according to the system they followed. Next week he played, then set her to prepare, the Nocturne in E Flat together with the more difficult Étude in E.

When Hoffman played the Nocturne—he played it twice—it made Marian catch her breath.

In a garden, she thought, drawing towards evening, still warm from the sun, everywhere singing with elegance, the flowers, light rippling on an octagonal pool . . . it's Chopin whose works I shall play. They'll be something I can devote myself to, instead of meandering and yearning . . .

At home, when Lucy had helped her change from her walking dress, Marian went to her music room, and retrieved the Nocturne in E Flat that Herr Hoffman had played.

It's not a difficult key, she thought. Except for the last bar but two, all the left hand is the same, gentle movements with single notes by the little finger, in progressions, followed by pairs of inverted chords. They set the tone for the whole piece, must be done gently or they would intrude. And this last bar but two, it's a kind of trill but with alternating flats and naturals. It'll be difficult to get the right pressure starting softly, then crescendo, then diminuendo.

She moved the slider of her metronome; the quaver at 132, as marked—it was far too fast to start with. She practised the left-hand chords, then the right-hand, finally both. And in the next days again, and again.

Three months later, on a cold short day in January, she had a realization. The fire crackled in the grate of her room, but its sounds faded as she found herself playing the piece as she felt Chopin would himself have played, of course not really, but with the animation of his spirit. He was playing through her, his touch in her fingers.

For some brief minutes, with a definiteness she had never previously experienced, the Nocturne became part of herself. Its inner movements, subtle changes of tempi, then the slight hesitation in the twenty-sixth bar but, as the next marking indicated, continuing *sempre dolcissimo*—very sweetly—leading in towards the climax of right-hand chords extending those of the left *con forza*, followed by the long sequel as the right-hand trill led towards the re-established rhythm and the soft final sounds.

In that act, in a provincial town five hundred miles from Paris, Marian attained a state not previously known, as if she were outside herself, as if her spirit transcended the mundane, became one for a moment with those harmonies by which the heavens themselves are moved.

Could this be what she had longed for? Bordered by frigid waters of the North Sea there had arisen this acoustic island, on which the soul could rest.

She had not previously applied herself fully to the works of any one composer. Now she knew that she would learn all Chopin's music, this piece the first thread of a connection.

She arose. A warm feeling spread downwards from her neck, beneath the bodice of her dress. She walked to the window. In the mid of winter, dusk falling towards evening though clocks insisted it was afternoon, she gazed south, her eyes reaching no farther than a streak of cloud against a fading sky.

Thoughts flew to Paris: a salon, she faced Chopin. Their eyes met in recognition, scarcely a foot apart. I hadn't known that music could do that, she thought. The piece is inside me ... I won't always be able to do it. It is enough to know.

Marian returned to her piano, sat down, played a chord, then rose again. Is this enough, she thought, to keep it at bay? the loneliness ...

It was a week later that she first met Leggate, two days after that the competition was announced.

In the next months Marian had continued with the Nocturne, started to prepare the Étude in E Major for the competition. She prepared too the Prélude Number 15 because, in an account she had read, she found it had been written when Chopin and Mme Sand were in Majorca: music for a loved one.

From time to time, within the next months, the experience of being at one with the composer was regained. Now in the long days of the summer of 1847, two weeks before the competition, the Étude and the Prélude were prepared. Each phrase, each transition of mood, was embedded in fingers and sinews to emerge in the touch of the piano's keys.

Although Dr Leggate had been part of the constellation under which Marian's new aspiration had been born, in the months of winter and spring his image had receded. As the competition approached, with technical problems brought under control, her questions had been how one part would flow into the next, how the repeat played in the same way would nonetheless sound different after the second subject, how the manner of phrasing flowed onwards throughout each piece. Only when these were settled had she told her mother that she would enter the competition. Only then did the question become salient once more of the doctor, of how he might be affected by what he would hear. Only then did she feel bold enough to talk to Caroline about him again.

"Remember I told you that I met a Dr Leggate, when he came to see my father after Christmas?"

"I remember," said Caroline. "I thought at first you were going to take an interest in him. But you've not mentioned him since."

"I've decided to enter the piano competition again this year. He's one of the judges. I thought I might be able to impinge on his consciousness."

Caroline laughed. "Good Lord," she said, "are you becoming devious in your old age? You're always so direct, too much for me sometimes . . . and now here you are setting snares."

"Not a snare. Don't scold me," said Marian. "Please, please . . . I feel bad enough about it. But he and I don't move in the same circles, and I didn't know how else to go about it, and . . ."

"I thought you'd become interested in Warrinder again."

"No," said Marian. "No . . . when I read that Leggate was to be a judge of the piano competition, I decided to get up some new pieces for it, and I've not been thinking of much at all, except of concentrating on them. I put him out of my mind. I'm sorry. I feel ashamed at myself, for not having told you, and for . . . it's not a snare. But when he came to see my father earlier in the year he was just a doctor and I was just a patient's daughter. He probably didn't even notice me, but how else should I make an impression?"

"I'm surprised at you," said Caroline, who then giggled and said, "Now you've been planning this . . . what a lark!"

Marian laughed too. "It doesn't mean he'll notice me now," she said. "And what if Miss Galloway were to win, and he notices her instead? She's young, she's very beautiful, she plays well."

"Don't be ridiculous," said Caroline. "Everyone knows you're the best pianist in the town . . . But are you really interested in him? You mean you've changed your mind about marriage?"

"I don't know," said Marian. "I don't know. That's why I've tried not to think of him, and haven't talked to you, it would overturn everything, starting to think of that kind of thing . . ."

"It would be wonderful," said Caroline. "If there were someone you really liked . . . and for you to have been so coy with me."

She looked at Marian in mock remonstrance, and laughed affectionately again, apparently at the idea that her friend now was starting to behave in the way that people should.

"When I met him after Christmas, he seemed the sort of man I could like, and have as a friend—he thinks a good deal."

"You'd enjoy that . . . thinking."

"I feel lonely sometimes. I often imagine leaving my father's house."

Chapter Eleven

When he had arrived in Middlethorpe, Leggate had been pleased to find that the town thrived, and that its growing prosperity allowed him to develop the practice he bought. He would not have been able to get started so well, however, except for Newcombe, who always said he had too many patients on his list, and would pass over to Leggate every new applicant for his services.

Newcombe had published several articles in *The Lancet* on respiratory diseases, including one on microscopic examinations of the lungs in pneumonia. Newcombe's face had radiated pleasure on hearing that Leggate was engaged in research.

"My dear fellow. So you are to bring enlightenment?" His white hair had made him look older than his fifty-eight years, but he remained vigorous.

"I'm determined to do something on cholera," Leggate had said.

"You must join our society—the Medical and Scientific Society. You'll find it provincial, but you'll do us good."

When Leggate had come to know Newcombe better, they would compare experiences. "I worked for a physician in Edinburgh," said Leggate. "Very well connected. He had two immense jars of concoction that he kept in a room he called his dispensary. These medicines were famous throughout Morningside. In one jar the mixture was pink, for ailments above the waist. The other was pale green, for ailments below the waist."

"The acme of scientific medicine," said Newcombe.

Leggate caught his eye; both men fell into an uproar of laughter.

I don't need to go running around, joining societies, Leggate had said to himself. What's important is that I mustn't be hindered.

The word had spread that Leggate had trained in Paris and Edinburgh. He became favoured by a certain kind of female patient. He began to reflect on his effect on Mrs Reckitt, or Miss Rank, or on Jane, or Elizabeth.

Solemn exchanges would occur. "It's my lumbago, doctor. It's usually better by May, but not this year. It makes my life a trial, and my husband makes it no better. And you know, when the time of the month comes, you know what I mean . . ."

"How do you mean, Mrs Reckitt, your husband makes it worse?"

"With his insistence, doctor."

The matters of which these women spoke were close to their hearts. He attended to such speech as if, at that moment, nothing else in the world mattered, and when he later reflected upon such consultations, he would look downwards, and would press three fingers to his brow, and shake his head.

It does no harm to listen, he thought. Remedies, physic, they're all largely nonsense, but for a person to be taken seriously is substance, so that by listening to them for half an hour, it makes a difference to them. It's not a burden, except for the time it occupies, it almost feels worthwhile . . .

Fortunate for our profession that ignorance is the least of our troubles, he thought. I should write a placard, and put it in front of my house. "The following is the only significant discovery on the medical treatment of disease. It was made by an Egyptian doctor more than three thousand years ago: 'The patient likes the doctor to pay attention to what he says, even more than to be cured.'"

Is this truth, or the worst kind of quackery? Leggate sighed, reflecting once again on the Egyptian adage, as he returned from yet another female patient. Mustn't think about it now, the important thing is to understand the patterns. To act effectively one must understand what one acts upon.

Since his best thoughts came when supine, Leggate had his housekeeper heat water in the copper, so that he could bathe after visits on Tuesdays and Thursdays, and luxuriate in the hot water.

Just two theories, he thought as his mind floated free of patients. Theory of contagion that a poison is spread by touching the afflicted person, or by inhaling air around that person. Contagion is what Wakley believes. Then there's the theory of the miasma, not from persons but from places, from filth as Southwood Smith and Chadwick have it.

Leggate leaned back, in amniotic warmth. He considered the compromise theory of Wilson: contingent contagion—the power of contagion but contingent on circumstance, of atmosphere, or filth, or some such... what would we deduce from that? In all theories something affects the victim, there must be some distinct thing. What? What thing?

At the back of the house in the scullery, a narrow beam of light from the westward sun slanted through the small window, illuminating in its beam bright specks that oscillated in the air.

A poison could be suspended in the air—it was a realization—like the motes in the sunbeam! he thought. I must have seen these specks in the sunlight a dozen times, I always just thought "a trick of the light," never realized, but they must be present all the time, usually invisible—that's their significance.

He was excited now: it's the beam that shines in a certain way to illuminate them, he thought, but the specks must be there all the time, some are poisonous, tiny particles of poison, powerful—because when the disease is prevailing the minutest quantity will suffice. Minute but insensible. Not quite insensible; the miasma can be smelled, and can be seen in a sunbeam...

Leggate's mind floated. Then... an implication; he jumped up, splashing water on the floor, forgetting that the purpose of bathing is to wash.

"That's right," he said out loud. "And therefore..."

He dressed swiftly, went to his study, and wrote humid thoughts at the beginning of a new notebook, *Cholera, Propagation and Physiology, Vol. III.* He concluded:

Can be smelled, and in the right conditions they can be seen too! Tiny motes in the air, some could be poison, some is other stuff, in a sunbeam. The different smells are the different kinds of substance; so if I locate places with different kinds of stench and some correspond to the concentrations of deaths, I can draw these species of stench on my maps. When the disease comes, it activates them. My maps will be the key, though previously I did not see exactly how, I shall practise detecting the different kinds of stink, then inscribe their areas on tracing paper, and overlay them on the map of the deaths, to see if any one kind of smell matches the high concentrations of death. The smells provide the contingency, when the activating contagion of the disease arrives.

Leggate wrote about the implications of this new realization, and formed a plan.

I shall walk about generally, he thought, learn to distinguish different smells. Go where deaths were most dense, start tomorrow. Practise discriminating between smells, give each a name, go systematically to make a map of odours . . .

"Dr Leggate." Mrs Huggins was knocking at his door.

"Come in."

"You've not eaten your dinner again, sir. A nice bit of beef."

"I'm sorry, Mrs Huggins."

"When I called, you said you was coming right away."

"I did?"

"It's been on the table an hour. I'll have to throw it away."

"Don't do that, Mrs Huggins, I'll eat it—I'm sorry that you went to that trouble."

"You can't eat it now, it's all congealed. I'll get you a piece of cheese."

Leggate was not sure why he allowed Mrs Huggins to talk like this. I've let her become too familiar, he thought. But he liked the sense that she cared about him enough to scold him, and did not just do mechanically what she was paid for.

"I was caught up in a problem," he said. "I appreciate your cooking. Please forgive me."

"I don't know how you can look after others if you won't look after yourself."

Chapter Twelve

Coming home from a visit, Marian, in her carriage, passed a group of men returning from the oil-seed mill.

Each one barely noticeable, thought Marian, just a stooped, ragged, grimy creature. Yet an immortal soul. A husband who comes close to his wife in marriage, a father of children, a son of a mother, once a tiny baby enclosed in her arms. Each with his own memories, loves, hatreds. Marriage perhaps makes their lives supportable, on the journey towards the grave. A dozen such souls. Each a centre of conscious life, looking out, regarding others, now regarding my father's carriage, with me in it.

Would marriage make my life supportable? Marian thought. It is supportable, I suppose. It is. But it lacks a centre. Each life needs a centre. I let Warrinder divert me too much. What if Dr Leggate . . .

When she reached home, Marian gathered up, as she did on Tuesdays and Thursdays, the bread that had been baked, with whatever else cook had put aside. With Lucy, she set out for her parish visits, to old Mrs Lorrins, to Mr Malloy, the tracklayer whose leg had been crushed by a railway sleeper, whose wife took in washing to help support their eight children, to Miss Teap who looked after an aging mother.

Should I be Miss Brooks, looking after my aging mother? thought Marian. Hard to imagine Mother needing any looking after. Married? A doctor's wife? What could be more conventional? But he thinks for himself—the very opposite of convention, something exciting in it, opposite too to Warrinder, whose forte is somehow putting into words what everyone likes to think.

But marriage? Would that be a centre for a life?

Although Marian had not seen or heard of Leggate in her circles since their meeting in January, Warrinder had seemed to be everywhere. Marian found herself, against her better judgement, following his progress. She wondered what the effects might be of his making the investigation of Marie Watson's murder a cause. Not just the performance of the police, but his own, would be subjected to the glare of public attention.

When Warrinder saw her in the street, a few days before the piano competition, when he offered to accompany her on her errand, Marian was given an excited monologue.

"We've been using the description that Millie, one of the women in the lodging house, gave of a man who was with Marie Watson on the night she was killed," he said.

Marian noticed her maid, Lucy, discreetly drop back so as not to overhear.

She assumes we're talking of something intimate, thought Marian.

"We don't have much," continued Warrinder, "but it may be enough, a description of the man, plumpish, perhaps five-foot-six, sandy hair, moustache, side whiskers, clean shaven, well dressed, a London accent. We thought there can't be so many fitting this description, not who consort with . . ."

Marian saw Warrinder look at her, as she kept pace beside him, to see if she would be offended. She looked squarely back at him.

"Anyway, my friend in the police, who's in charge of the case, brought in a couple of fellows who fitted the description, and asked Millie to look them over. It wasn't them."

Marian kept in step, and glanced at him, to see if he were waiting for her to say something. He merely needed her look of encouragement.

"'Why don't I question these women again, see if I can't squeeze something out of them?' I said. 'Give me a few days,' I said, and Millie . . .'"

Marian wondered why he paused, wondered whether that was not a blush on his neck. What now, my charming Millie? she wondered.

"Millie's been very helpful, so we're going to places where we're likely to spot the fellow. Not here, you see, but in Leeds," continued Warrinder. "Millie remembered Marie Watson saying that's where he's from. We've not let that information out yet, but I think going to likely places there will be the way to catch him. You'll see if you follow the story in the paper."

CLOSING ON PERPETRATOR

OF SACRIFICIAL KILLING

From our own correspondent

Working in collaboration with the Head of the Department of Detection at the Middlethorpe Police Station,
a gentleman who prefers to remain anonymous is taking
steps that may lead to discovering the identity of the
killer of Marie Watson, poisoned with strychnine, who
was lately found before the west door of St Michael's.
The dead woman lived with a number of other women
in a notorious house in Friars Walk. Not only do the
women in this house know the comings and goings of
some of this town's more illustrious citizens . . .

Marian read the article. Warrinder—"correspondent," "gentleman who
prefers to remain anonymous," she thought. He might as well go about in a
cart with his name in foot-high letters.

Interest in the case multiplied. Warrinder convinced his uncle to print
more copies of the newspaper than usual. All were sold.

As if some noisy contraption were echoing round the town, news about
a man from Leeds reverberated, rounding corners, penetrating doors. A man
of standing! The murder of the prostitute took yet stronger hold on public
imagination.

Rutting and killing, thought Marian. Lord, what beasts men are.

II.
Romanza

Competing and Courting

It is from you that I receive all life, on whom I am wholly dependent. Like a slave, I should often like to follow you from afar at a distance, and await your slightest bidding. Ah! let me say it once more, come what may — I will whisper it even to whosoever closes my eyes, "One alone has ruled my life completely, drawn me into her innermost being, and it is she that I have ever honoured and loved above all."

Letter of December 1838 from the Romantic composer Robert Schumann to the pianist Clara Wieck, written during an engagement that was forbidden by Clara's father. A court ruled in favour of the couple, and Clara Wieck became Clara Schumann one day before her twenty-first birthday.

Chapter Thirteen

On the 21st of June, still intent on smells, Leggate shrank from having to judge a piano competition.

Young ladies tinkling on an instrument they're unlikely to comprehend, he thought. Distraction... I should have refused.

He arrived at the Theatre Royal to discover in the entrance hall the originator of the competition, Mr Brown, and accompanied him to the box reserved for the judges. He was introduced to Lord Andervilliers, the third judge. They arranged themselves, and Leggate sensed the audience's attention upon them.

"I rather like this auditorium," remarked Lord Andervilliers, who was known for his interest in this whole area of the county.

"Musical accomplishment not at all bad here," said Mr Brown.

After a little time the performances began. The first two pieces were pleasant, the pianists equal to what they had chosen, the audience appreciative.

Every day Leggate drove himself; his housekeeper woke him at half past five. If he were called out during the night, the next day he would drive himself forward nonetheless. Last night he had been called out. With the playing of the third competitor his eyelids were drawn earthwards with a force stronger than gravity. His head nodded.

He started awake to the sound of applause. His programme dropped to the floor. He made inconsequential movements to retrieve it, and when he looked up again, towards the stage, the performer had left. He drew the fingers of both hands across his closed eyelids, as if smoothing on some ointment. Then a perceptible change: murmuring among the audience, an apparition, a young woman, on the tall side with a light figure, wearing a beige gown trimmed with gold brocade, glided across the stage from beneath the judges' box. His eyes were full upon her. She turned, curtseyed to the audience, seated herself at the piano.

"The next performer is Miss Brooks, who will play an Étude in the key of E by Frédéric Chopin," announced the Master of Ceremonies.

Miss Brooks glanced towards the judges. Leggate's eyes met hers.

Who can she be? he wondered.

She paused to await a quietening among the audience . . . She began to play.

The piece began gently, with a melody that stopped the breath. Within the first few bars Leggate was fully awake. He could not have explained it: no longer in a provincial theatre, he began to feel as if a great loss overwhelmed him, involuntary tears welled in his eyes. He did not try to stanch them. Must find my handkerchief, he thought.

The mood of the music changed. Exuberance? he wondered, or chaos following great loss?

Now it seemed that every note on the piano was brought into action. Through damp eyes Leggate looked at the performer with wonderment. How does the human anatomy produce such movements? he thought.

Then, as the bravura passage finished, a linking passage promised hope, which gave way again to the grand melody with which the piece had begun.

The same as the beginning, he thought, the same notes, now utterly different, still sad but inexplicably, as if with a knowledge and understanding of possibility that's grown out of the middle passage, the loss accepted, surmounted.

The music ended. Leggate found his handkerchief, coughed, was able to wipe his eyes, he hoped without attracting notice, for others' eyes still were on the performer. It had been a shorter piece than those played by other competitors. The young lady had risen, walked towards the front of the stage, curtseyed.

During the applause Lord Andervilliers leaned towards Leggate. "I'd be surprised if the others beat that," he said. "I met Chopin when he was in London, don't you know. Opinion is that he is the finest musician in Europe. Think I agree, though some like Liszt. Plucky of the girl to play that. Not an easy piece."

Leggate looked at his programme to see who the next performer was to be, then looked up to see Miss Brooks make her journey back towards the wings, beneath the box. Halfway across she raised her eyes. They met his. Although she made no explicit greeting, Leggate involuntarily nodded slightly, as if in acknowledgement.

The last four performers played. For neither audience nor judges did they create any comparable impression. When the judges conferred, they agreed at once that Miss Brooks was first. Their difficulties came in separating second from third. His Lordship proposed that Miss Galloway should be second, since the fingering for the piece she had played required skill—so the results were decided.

The custom was for the first three in the competition to be announced by the Master of Ceremonies in reverse order, then the one judged first would play a final piece before the interval. Miss Brooks appeared once more, approached the footlights, curtseyed, and said: "I shall play another piece by the French-Polish pianist and composer, M. Frédéric Chopin, his Prélude Number 15, in D flat."

The piece opened in tranquillity, a secluded garden. The mood gave way to a set of repeated, mournful chords in the left hand, then a crescendo, then the same mournful chords once more.

As if some sorceress were at work, he thought, their meaning changed, giving the sense of being able to rise from an abyss, she has transformed these chords into a grandeur...a kind of triumph. The second time she has accomplished such a metamorphosis. Is that Chopin, or is it she?

Then the piece returned to tranquillity, the opening mood now also transformed, then dissolving into quiet hopefulness. The audience had been held suspended. Then there came a rush of applause.

She chose this not as the piece to be judged, Leggate thought, but as the performance for which she wished our attention—not a competitor but an artist.

"That woman is capable of great feeling," murmured Andervilliers to Leggate.

After the final piece the judges were to go onstage for the presentation. "Follow me," said Mr Brown. Down some stairs they followed him, along a short corridor, through a door concealed in wooden panelling, then into the actors' regions. With Evette, Leggate had often been backstage; he was unable to quell an uprush of recollection.

On stage now, Miss Brooks was presented. She shook hands with Mr Brown, who congratulated her. Now she came to stand before Leggate, with brown eyes regarding him as if in recognition.

He took her right hand in his, the conscious touch of another human creature. He was surprised her hand was bare ... Of course, he thought, in these circumstances, no gloves.

"My congratulations, well deserved indeed," he said.

"Thank you, sir," she replied, keeping his eye longer than expected. She passed on to curtsey to Lord Andervilliers, shook his hand, received a scroll.

Holding this she moved again towards the front of the stage. More applause, another curtsey. Leggate noticed movement behind him as men prepared to bring down the curtain. Now something occurred that the organizers had not anticipated. First one, then more people in the audience rose to their feet and shouted: "Encore." There was a renewed crescendo of applause, more stood. The Master of Ceremonies moved towards Miss Brooks and exchanged a few words; he stepped forward and raised his hand to quieten the audience.

"As an encore," he said, "Miss Brooks will play a final piece before the interval. This one also by Frédéric Chopin, his Nocturne in E Flat."

The judges retired to stage left. "Better wait here," said Mr Brown.

Again Miss Brooks played, and Leggate found himself wondering: Could she have prepared this, too, in case an encore were requested? He thought of himself once more in the theatre, place of artifice, where he would await Evette after the play, talk with stage hands, accepted into the secret fraternity who lived on this other side of the great curtain. The notes of the nocturne began. He had heard it played by Chopin himself in Paris—its honeysuckle sweetness transported him back to that time.

In love, he thought, that's what it means ...

All at once it was ended. Applause was renewed, Miss Brooks again rose, again approached the front of the stage to curtsey, stepped back a pace, and the curtain fell.

"Quite charming," said His Lordship. "Crowd pleaser, of course, not so fine as the other two. I'm glad she didn't play that first."

Mr Brown was saying: "Follow me. Refreshments are served. The mayor and aldermen, and . . ." His words became inaudible.

Leggate found himself thinking: to look into those eyes again . . .

In the refreshment room Lord Andervilliers confided to Leggate, "I'm not keen on this part. We did right though, wouldn't you say? That young woman could acquit herself on the concert stage. I have to say, though I'm sure I shouldn't, that I'm surprised there was anyone as good. Her choice of the Chopin was shrewd. Some parts technically demanding can't be well known to this audience. Here she comes now."

Marian approached with her mother and father.

"Congratulations, Miss Brooks."

"Thank you, m'lord," replied Miss Brooks. "You are most gracious. May I present my father, Mr Brooks, Rector of Swanscombe, and Mrs Brooks."

"How d'y'do. Pleased to meet you."

"How do you do, m'lord."

"Dr Leggate—Mr and Mrs Brooks."

"How do you do."

"Dr Leggate attended you in January, when Dr Newcombe was away," said Miss Brooks to her father.

"Indeed," said Mr Brooks. "Then I should thank you, doctor. You see I am quite recovered."

"I am pleased." Leggate looked towards Mr Brooks, to catch his eye in a friendly way, but received no acknowledgement.

I met her before! he thought. Of course . . . she said, "Shouldn't you act on the disease?" and we argued about remedies.

"You have a most accomplished and charming daughter," His Lordship said. "Her choice of pieces very fine, too. Chopin just right. Pleasing to hear composition in the French style. My family was French originally."

"Indeed, sir," Mr Brooks replied.

"Pray excuse us," His Lordship said. "The judges have a further matter to transact before the next part of the programme." He touched Leggate's arm to motion him away.

"I didn't know we had anything more to do."

Leggate, in some agitation, was thinking over the forgotten encounter with Miss Brooks: it was when I was locum for Newcombe, he thought. The same woman! I wouldn't have known; she's got up for the stage, I suppose, but I should've said something to acknowledge we'd met.

"I wanted to get away," said Andervilliers. "Don't ever know what to say in that sort of situation, they just stare at me. If I didn't make a move we'd be stuck there for ever, like so many damn statues. That young woman was something though? Handsome too—and her interpretation. First rate. Composer would have approved."

"I thought she was extraordinary," Leggate said. "I heard Chopin in Paris," he added. "He played that E Flat Nocturne. I agree with you that he's the foremost musician of our age, a successor to Mozart and Beethoven. The nocturne's not so impressive as the other pieces, but she avoided the temptation of too much rubato, and brought out its simplicity and grace."

"You don't say." Lord Andervilliers looked at Leggate. "You've heard Chopin? So, you've travelled. Where are you from?"

"I grew up in Dulwich, to the south of London, but my family's from Dorset."

"You must come over to the Hall, for one of our musical evenings. Give me your card. I'll make sure my secretary puts you on our list."

Leggate produced a card.

"You'll get an invitation," Andervilliers said.

A man came up to Lord Andervilliers and greeted him. Leggate was introduced, but when two more men came to join the group, and seeing the conversation would proceed without him, Leggate slipped away, and found himself drawn inexorably back towards Miss Brooks, who stood next to her mother and father as they conversed with another couple.

Miss Brooks looked up as Leggate approached. "Your judicial duties are complete?"

"Entirely," said Leggate. "Never was a judgement more correct."

Marian smiled, as if to indicate that this gallantry was unnecessary.

"When you perform like this," Leggate said . . . he moved sideways to allow some people past, and with the readjustment moved slightly away from Miss

Brooks's family group so that she too stepped aside with him, "how do you think of the music? As something you need to get right, or does it move you, in something like the way the audience is moved?"

"For me getting it right is done a good long time before," she said. "I think of it mostly as a kind of poetry, rhythms and metres of course, but the equivalents of alliterations and images, though without words."

"I used to read a good deal of poetry, mainly seventeenth century."

"Donne, Marvell, people like that?"

"Exactly," he said.

An educated woman! he thought.

"But no longer?" she continued.

"Science now, almost exclusively, but mixed with a little music."

"So you've constricted your life?" she asked. She sounded surprised, and the directness of her question arrested him. He looked into her face, saw her even features with high cheekbones, saw brown eyes looking back at him. The bell rang for the end of the interval.

"Perhaps I have," he said. "Perhaps I shouldn't have done."

"I must go," she said. "Good evening to you, and thank you so much for your good opinion of my piano playing."

That evening Leggate wrote:

> We awarded first place to Miss Brooks, who played a piece by Frédéric Chopin, an étude far superior to anything played by the other competitors, and then as the winner she played a difficult prélude, and afterwards a nocturne; and, without exaggeration, it is not since Chopin himself that I have heard such fine touch, with intonation full of meaning, and the melody emerging as from a hidden depth.

Leggate paused. This was a candid diary. Even so, perhaps he was saying too much.

> If I were to marry it would be someone like that. Life can be tedious here, with patients and researches, but if in the evenings I could listen to

sublime music, then to bed. She looked at me, I thought, meaningfully, might that signify something? And to think I didn't acknowledge that we had met, but I know nothing about her, she must be already promised, but perhaps someone like her . . . no one else could play like that, not in Middlethorpe.

Halliwell is happily married, in the evenings and on Sundays he is refreshed, among his family, that is the pattern. Time for work would be lost, but there would be compensation because I would work better, and it would bear me through disappointment, and ease the plodding of my route, for is there not a woman behind every man who accomplishes any great task? And I could do it for her, all the better for her.

Chapter Fourteen

During the night after the concert Marian did not sleep well. As she lay in bed her thoughts returned to the theatre.

Had to prepare two pieces, she thought, we all did, but an encore, what a thing to happen!

"Ooch." A sound escaped her lips.

So of course, she thought, mentally rehearsing the events, I had to play the Nocturne, nothing else would've been any good, though I hadn't played it for two weeks... or for three, it must have sounded rusty, rusty, rusty. I'd been concentrating on the others... when I made that mistake in the Étude, I thought I won't be able to recover, I'll have to stop and get started on the phrase again, to get onto the sequence again, so that it would be obvious to everyone. But somehow I played through it, the judges didn't view it harshly. My aunt said that everyone wanted to hear me, that people had liked my piano playing in the competitions when I was younger, that since I'd been to Germany I had become something of a personage, one who goes abroad, and I'd persuaded Mendelssohn to come, so that's why they were so welcoming to me. Touching, how kind they are, what if I'd let them down? And then Dr Leggate... The second piece, at least, went well. Not like Clara Schumann, of course, not even as well as I sometimes manage at home, but that's when I can forget myself, but well in the sense that I played almost as I would like, with my limitations.

And then when I spoke to Father, she thought, and asked if he weren't pleased that his piano-playing daughter had won the competition, all he could say was, "Well, of course you'd win, you don't think I've been paying all that money all those years for nothing, but why all you'll play nowadays is that French frippery I can't make out."

At breakfast Mrs Brooks said to her daughter, "I was so pleased to see you play again in the competition. It made me proud."

"I am glad, Mamma." A self-deprecating smile, then, after a pause: "Father seemed cool to Dr Leggate. I thought he was very thorough when Father was ill in January. He is not a man to take exception to."

Without looking up from his novel, William said, "Marian's taken a shine to the doctor."

"You should guard against such ideas, William," said Marian. "They will disturb your idleness."

"Your father prefers Dr Newcombe. He doesn't like change," said Mrs Brooks. "He said you and the new doctor argued at his sickbed. He thought it improper."

Marian felt a surge of exasperation. Why does Mother say this to me now? she thought. She decided against making a reply.

"My sister's very satisfied with Dr Leggate."

Marian saw that her mother realized she was displeased, and was attempting reconciliation by way of the favourite aunt.

"She says he's very presentable," continued Mrs Brooks. "That he doesn't believe in the lowering treatment."

Marian still did not reply. Her mother continued, "They say Dr Leggate is of good family."

"Dr Newcombe says they're from Dorset," replied Marian.

"Your aunt's still not well," continued Mrs Brooks. "Come upstairs, and play that last piece for me again."

"I shall go and see my aunt again today," said Marian. "Some people think that last piece shallow," she said, still aggrieved with her mother.

"Play another piece then, whichever you prefer."

As they left the dining room Mrs Brooks said, "Let cook know when you've finished breakfast, William. So that she can clear away."

In her music room, Marian allowed herself to feel mollified by her mother's tone. She was glad to have escaped William's satirical commentary.

"Do you think Mrs Larkin would invite Dr Leggate to her dance?" she said. "He's not been out much in Middlethorpe. It might do him good."

"I'll have a word with her," Mrs Brooks replied. "A single man is always welcome at a dance."

Marian sat down at her piano. She thought of how she was said to take after her mother, who would still sing when asked, and be congratulated by the gentlemen...

Why does she goad me about arguing with Dr Leggate when Father was ill? she thought, her irritation returning. Does she hold something against him, or just wish to make some point? Then why not say so?

She searched through her music at unnecessary length. She played three times the passage in which she had made the mistake in the competition— a slight fault unlikely to have been noticed by anyone who did not know the piece intimately—before at last she played for her mother the whole piece, perfectly.

Marian had visited her aunt twice since she had been ill.

If I go again today, she thought, in the afternoon, perhaps on the off chance I might be there when he calls.

"I am sorry you're still not better," she said to her aunt, as she sat at the bedside.

"I'm improving. I hear you won the competition, I'm sorry I couldn't hear you."

"I felt as if I might be able to give performances, real ones. It's so heartening when people applaud, when they are appreciative. It makes all the effort seem worthwhile."

Marian saw her aunt looking at her, appraising, making Marian wonder whether her aunt would be proud if she were to become a real performer.

"Would you like to talk?" said Marian. "Or shall I read you something invigorating—a mental tonic?"

"I feel too passive to talk," her aunt said. "Some Jane Austen. She's engaging but not too invigorating. You know where they are, just choose something. With the great authors, you can just dip in, and even a short piece is full of meaning."

Marian walked to her aunt's bookshelves, selected a volume: *Persuasion*.

"Open it at random, and see if there isn't something that sets one thinking."

Marian seated herself, then did as her aunt suggested; she began in the middle of a sentence a little way into the first chapter: "... but Anne, with an elegance of mind and sweetness of character, which must have placed her high with any people of real understanding, was nobody with either father or sister..."

The words captured Marian's thoughts as she read. Towards the end of the chapter she heard her aunt's breaths, audible as sleep came over her. Marian closed the book, walked to the window, glanced up the street.

She resumed her position on the chair beside her aunt's bed. Still sitting, she did not know how much later, she heard the bell downstairs. Glancing towards her aunt, she went to the door. By the conduit of the stairwell she heard sounds of a man entering the hall. She returned to shake her aunt gently. "The doctor's here," she said.

"I must have fallen asleep," said her aunt.

"I'll ask him to wait a minute."

"I'm all right. I'm awake now. What did you think about my proposition that one can read a passage at random from the great writers?"

"Within a paragraph one's right in the middle of it, the words are enough to make one want to read them again."

"Persuasion," said her aunt. "I sometimes worry that I've persuaded you too much."

Footsteps were heard at the top of the stairs.

"The doctor, ma'am," the maid announced.

"Ask him to come in."

Marian placed the book on a small table beside her aunt's bed, rose from the chair.

"Miss Mitchell," said Dr Leggate. "Miss Brooks!" An involuntary elevation of his eyebrows lasted a small part of a second before his face relaxed into the warmest of smiles as his eyes met hers. It was a sign Marian had hoped to observe.

"Dr Leggate," she replied. She lowered her eyes, as if acknowledging some shared secret.

"You know, Aunt, that the doctor was a judge in the piano competition?" said Marian. "He was very generous to me."

"Entirely deserved," said Leggate.

"I shall go now," said Marian to her aunt. "I hope you'll be well soon. You're much better than last week."

"I think I am. Thank you for coming. Tell your mother I'd like her company."

"Goodbye, then. I have several things to do. I must go home to Sutcliffe Street first. I'll tell Mamma you'd like to see her."

Marian called for her maid, and was soon in the street. She walked slowly, for nothing urgent needed her attention that afternoon.

Chapter Fifteen

Marriage, thought Leggate. It would make it all worthwhile. Less time for research, of course, but . . . music in the house, she would be the sort of woman . . . Mustn't think about that now.

Leggate had no further patients that day. He could be seen sniffing at putrefying piles of ordure, at piggeries of gut-heaving stench, then around the building where the boiling of whale oil gave off a characteristic smell that wafted through the town.

She must live in one of the better parts, he thought, better because fewer stinks, less susceptible, Sutcliffe Street she mentioned, yes, that's definitely in one of the better parts. Some rise above the filth, into the realms of music . . .

A melody from Miss Brooks's performance echoed in his mind; it seemed incongruous here in the streets.

Extraordinary to meet her there at her aunt's, he thought. Extraordinary coincidence. Miss Mitchell is educated, no doubt she has influenced her niece, and so in two days, from barely meeting her, then forgetting—how could I have done that?—I see her twice. Something extraordinary about her, not since Paris . . . Can't think about that now.

He resumed sniffing. He peered through archways into dank courts. In each a single noisome privy was used by several dozen people, each surrounded by its swarm of disgusting flies, each broken by the spades of the coarse fellows who infrequently removed the night soil.

He could not discriminate smells as well as he hoped. Against that, he formed a new idea, that the cesspools, piles of rotting matter, heaps of the excrement of horse and human, or the shiny, foetid, open drains—principal causes of some of the most disgusting smells—must have been in place for fifteen years. When he enquired, he found that most had indeed been in their current positions for as long as anyone remembered, part of life. He began to map on tracing paper the locations of these repellent monuments.

But returning to a place previously visited, he could not swear that yesterday's smell was the same as today's, so he decided to make just a coarse discrimination. Some smells he labelled "O," for ordure, while those emanating from a trade he labelled "T." Patterns began to form: Os were ubiquitous, but most reliably around the cesspools, drains, and heaps of excremental rubbish, and then all around these were clouds of flies.

Following a report by a Committee of the Middlethorpe Medical and Scientific Society which described the appalling sanitary conditions and inadequate drainage of the town, some ineffectual removals of rubbish heaps, described as nuisances, had begun, but if a nuisance were removed, in two days it returned.

Leggate was grudging towards the sanitary report. It was thorough, it made sensible recommendations, but he disliked thinking of it.

I should have been asked to serve on that Committee, he thought. Must get Newcombe to introduce me to that Society, as he promised, but it's good that most nuisances are still *in situ*. They may be vital ...

That evening he went to visit Newcombe.

"You said you would introduce me to the Medical and Scientific Society?"

"I would have done so long ago. You seemed lukewarm."

"I would like to ... but not immediately, when my maps are finished, then I can report my findings to a meeting."

"You let me know."

"I oscillate between thinking I'm making progress and a sense of impossibility," said Leggate. "Just now I feel I am on to something, about miasmata..."

"I'd be interested to hear."

"It's not quite there yet. You probably didn't know, I was judge of the piano competition."

"I knew the music festival was on."

"One performer was remarkable. Miss Brooks, whose family are your patients, I visited the father when I was locum for you, earlier in the year."

"She's much admired. She went to study in Germany."

"She's very accomplished."

The summer evening was warm; the light had ebbed, and the gas lights had been lit, as Leggate walked towards the docks and towards home.

Should've asked Newcombe more, he thought. Couldn't go further without betraying too much interest. How can I find out? Don't move in the same circles. Don't move in any circles really, only the Halliwells and Newcombe, and the choir a bit. Perhaps ask Margaret's advice, how to find out more about her...

Leggate turned a corner. As he entered the dock area, a repellent stink assaulted him.

That's what I must think about—smells, he thought. Where were we? There is some correspondence among nuisances, smells of ordure, and the deaths of 1832, but it's like poverty, in some places where you expect correspondence deaths were sparse. Should I find another method?... or will the true causes always be beyond our senses? No: there may be a correspondence, but my maps are fifteen years too late, so correspondence can only be approximate. Some midden heaps and cesspools are exactly where they were, but some must have shifted... Await the next epidemic, because my method's in place, and Miss Brooks, what would she think of the discoverer of the cause of cholera?

Chapter Sixteen

In his study, Leggate lit two candles. He leafed through twenty-eight note-book pages in which he had summarized everything known about the disease.

Set out with utmost clarity, he thought. Miasma, contagion, combination of the two.

He rose from his chair, looked through the twilight towards the river. He saw the dockyard, moored ships on the river, quiet now. Reflected in the window, among these outer objects, his candles illuminated papers on his table, the world outside mingling with the candles and paper within.

That's the problem, he thought. How to see just what's there, without obscuration by what's within. Or is it the opposite? To detach from sensory immediacy of sights and smells, to infer a process beyond the senses.

He sat again at his table. I've certainly made an advance, he said to himself. Separated refractions from reflections.

On a large separate paper he had compiled a table: "Summary of facts, theories and speculations." He hoped facts might delineate a meaningful figure, the reflections of irrelevancy fade to background. There were connections—two facts were salient. He had entered them on a fresh page, and given each a number.

1. The first place the disease strikes is the mucous membrane and capillaries of the alimentary tract, i.e. large and small intestines and usually the gastric cavity. We know this because the first results are diarrhoea, and usually vomiting. In other diseases, by contrast, the first manifestation is typically a general fever affecting the whole somatic region.
2. Evacuations are almost entirely fluid, the so-called rice-water stools. We know the patient loses this fluid from the blood, because Magendie and I showed it when we injected mercury, and saw that droplets traversed the walls of the blood vessels into the lumen of the gut. We even saw how

the evacuations are created by injecting fluid, and seeing how particles of lining of the gut were dislodged.

Conclusion: the disease attacks first the alimentary canal, especially its blood vessels, causing permeability, and hence the loss of this fluid as diarrhoea.

"That's right," said Leggate aloud. "Then . . ."

Leggate thought he had not seen this conclusion written by any other person, at least not in such succinct form.

One could be satisfied with that advance alone, he thought, compared to all the twiddle-twaddle written about the disease. A worthy suitor, my income's adequate, and I'll join the Medical Society, have to enter a bit more into the life of the town, I suppose, become known for my researches, but the word'll spread of the discovery. The trouble is, this isn't much farther than the point I reached fifteen years ago.

He rose from his chair exasperated, paced, lay on his couch. That look she gave me at her aunt's, he thought, it wasn't an ordinary look, it was meaningful.

He rose from the couch, and resumed pacing. Not now, he admonished himself, mustn't lose momentum. All previous theories hold that something is inhaled from invalidated air—the air, pestilential vapours, that sort of thing. If correct, why then doesn't it first attack the lungs, as in tuberculosis or pneumonia?

He sat once more at his table.

"First attacks not the lungs but the alimentary canal," he wrote—"highly significant."

He underlined the last two words, looked again at this conclusion. Although he had written almost the same thing on the previous page, it had a comforting quality. It was precise. Better still, it was complete.

He rose once more from his chair, went to lie on his couch, gazed at the ceiling, in the wavering candlelight.

Title for paper, he thought, in *The Medical Gazette*: "Cholera, a disease not of the whole body, but of the gut." No, gut's wrong—"of the alimentary tract."

Again, however, he suffered a familiar discouraging thought. I should have advanced in fifteen years, perhaps by now it's known, though maybe not, just as probable that that's how discoveries are made, by one person seeing something with utter distinctness among obscuring details.

Leggate started; he had dozed. Friars Walk, maps, smells, nuisances, Chopin's nocturne, permeability of blood vessels, all jumbled in his mind, as if tending towards some connection among them all.

He shook himself wide awake. Smells, the gut, permeability... he thought. If it were the gut, the disease comes from what people eat. Particles of poison, specks in the air, and whatever-it-is doesn't affect people if they just inhale the air. I walked about when the disease prevailed in Sunderland, even more in Paris. But if it affects not the lungs but the gut, then it must come from food, so it's those who eat where the smells prevail, only those people contract the disease! That's it—it comes by the food. It doesn't affect everyone, because some have stronger stomachs.

Leggate's thoughts now moved purposefully forward. Are those specks the particles of poison in the air? But this is the miasma theory, poison in specific places, but both the miasma theory and the theory of contagion could work together—and here's the difference, not inhaled, but the poison gets from the air onto food, and is swallowed. This is the link.

He was excited now. "That's right!" he said aloud. "It really is."

What could such a poison be, he thought, that has such immense effect in such minuscule quantity? Follow this thought: a poison that's effective in infinitesimal quantities, a substance like strychnine or belladonna, but a thousand times more powerful. When the disease prevails again, I could scrape some of this poison from the dung heaps, where the smell is strong, buy a rabbit, freshly killed, apply it directly to the rabbit's intestine perhaps, or make a solution of it and bathe the intestine, that would be better, then repeat the experiment of injecting mercury into the mesenteric blood vessels. The poison would have its effect by causing permeability, and I could visualize the permeability with the mercury. The critical demonstration: from poison to permeability!

Leggate rose again from his couch. It was completely dark now, the candles were low. He turned a page of his notebook, wrote the satisfying conclusion:

Suspended in the air, in minute droplets when the disease prevails, there is a poison, innocuous when inhaled but permeates food so that when the food is ingested the poison exerts its effect by perforating the alimentary blood vessels.

"She'd be proud to know a discoverer." Leggate surprised himself by once more speaking out loud.

In the short days of winter his thoughts had turned slowly. If there were movement, it had been slight. Now as with steam at full pressure, when the engine room telegraph has signalled Full Ahead, his thoughts began to revolve with the satisfying breath and whirr of a well-tended engine.

Next day, he rose early as usual, he had slept only four hours, he wrote some more, took breakfast, fretted at having to attend patients.

Can the comfort of two or three be weighed against a good for all? he thought. He did attend, but he also thought, also recorded stench and nuisance in the town, also made new maps. He was excited by the idea that the answer was in some specific poison, that could be smelled and would float from midden heaps into the air, and then waft onto human food.

By midday *The Times* had arrived from London with news of new cholera outbreaks in Persia.

The wave is swelling, he thought. It'll traverse Russia, reach the Baltic, and then it'll be here. Methods, maps, I must have everything in place, my mind alert to every indication.

Within an hour of seeing his last patient next day, Leggate was again out sniffing, wrinkling his nose at stenches, and waving away in disgust the clouds of flies that attended every nuisance.

He continued to make plans. When the epidemic arrives, he thought, and I scrape some smelly poison from midden heaps and make aqueous extracts, and show permeability of blood vessels in rabbit's gut, it'll be an experiment of which even Magendie would be proud, so I must get in supplies of mercury, and syringes to inject it into the mesenteric arteries. Rabbits can be got any day from the market, so that needs no preparation and they keep them in cages there until they're sold, so I can ensure the animals are newly killed, but I mustn't neglect other aspects—in my article I must show that dung heaps mean deaths, first in 1832, and then in the coming epidemic.

In a small region of streets where deaths had been dense, near where an open sewer flowed, he made sketches for his map of smells. Then a heavy downpour started, washing the smells from the atmosphere. Leggate pushed his notebook beneath his coat to keep it dry, regarded the muddy water mingling with the droppings of horses in the street to make lazy rivers of filth. He picked his way back slowly towards his house, oblivious of the rain, thinking, thinking.

The poison, he thought, then, when it stops raining, these frightful flies. No muck without flies. Are the flies attracted to poison? Loathsome creatures. A poison forms on these heaps of muck, then flies walk on it...

He turned the corner into Fishersgate.

"Dr Leggate!"

"Amy—I didn't see you."

"You're in another world, sir. I was right in front of your eyes."

"I'm thinking of a problem."

Amy had an easy familiarity with Leggate. "Come under my umbrella," she said. "You're wet through. Someone should look after you, or you'll be under a cart."

"I'm still on the problem of cholera."

"I won't keep you. I just wanted to say that Connie and the baby are doing well, and we, Connie and Liza and me, we are going to bring up the baby together. Connie's calling her Florence, after a lady she liked, who she was in service with in London, so she'll be Florence Smith, a nice respectable name. We'll make sure she gets a better life."

"I wish you well. I would be honoured to be physician to this young person. I offer my services to your household free—there, that'll be my contribution. Why don't I come and see mother and child right now?"

"They'd like that, sir."

In the kitchen of Amy's house, Leggate accepted a towel to dry his face. They went upstairs. The baby girl was healthy, and the mother fully recovered from her delivery.

"It's good to see you both so well," Leggate said. "Amy will tell you, I'm appointed physician to the household. I will do what I can to help this young soul."

Leggate went downstairs, was in the hall when a knock came at the door. It was opened by the housekeeper. Mr Warrinder from *The Middlethorpe Examiner* was admitted, dripping water onto the floor from his coat. Leggate had met him soon after he had come to the town.

Warrinder arched his right eyebrow. "Dr Leggate is it not? Do I remember aright? So this is how we spend our afternoons."

"I bid you good day, sir," said Leggate.

"I am coming to interview these young women for the newspaper, to see if they can give me information about the murder of Marie Watson." He paused, and eyed Leggate. "And about the men who visit here."

"I trust your enquiries are fruitful."

With this Leggate walked directly towards the door, causing the new entrant to step aside to avoid a collision.

What an unpleasant, insinuating person, thought Leggate. Who ever does he think he is?

Next day Leggate had two calls, the second on the northern edge of town where the well-to-do had extended an enclave. Rather than coming directly home, he rode his horse along a path towards the river. He dismounted, then walked across stepping stones of a tributary, leading his horse splashing through the water. He sat on a fallen tree as he allowed his horse to drink and to eat the grass at the edge of the stream that flowed brown and thick with yesterday's rain. Insects buzzed amid buttercups, a contrast to town filth.

Something reminded him... What was it when Amy interrupted me yesterday? he thought.

The memory could not be recaptured. Leggate fell back on reiterating comforting truths, now etched in his brain. Poison identified by especial pungency, he thought, it forms on the filth, then wafts into the air as minute particles, some of which are specks of the poison that land on food.

A fly landed on his wrist, he flicked it away.

That was it, when Amy interrupted, he thought, flies land on the filth, they must get the poison on their legs … then they walk on meat that will be eaten. So that's how the poison reaches the gut. That's it. It's the flies … Bees spread pollen from plant to plant, flies are their sinister counterparts, propagating disease from stinking dung to food that will be eaten by people … That's it! It's not the air, that is what everyone got nowhere with, and if there's poison in the air, we should all be ill all the time, and with a gale of wind it would be transported from Persia in two days. So that can't be it. Instead it's much slower, more specific.

Leggate interrupted his horse's grazing, and started to lead it again. Flies would do it, he thought. Why didn't anybody think of that? It would explain the miasmata, correspondence with stink, but the poison isn't the odour itself. What the odour must do is to attract the flies in some way, so they land on the filth, which is always damp and mucky, and the flies then get poison on their legs as they wade about in it, and then they fly off and walk on our food.

He mounted his horse, reached the river, then started trotting the elderly animal along the towpath. He was excited, and wanted to stop some passer-by to explain his idea, perhaps he would find Amy to tell her, or go immediately to Halliwell's.

Communication to *The Medical Gazette*, he thought. That's the connection—from mode of communication to physiological affection.

At home, Leggate recorded the idea of flies in his notebook. Then he wrote a title.

ALIMENTARY LEAKAGE FROM BLOOD VESSELS AND
MEANS OF PROPAGATION OF CHOLERA MORBUS
JOHN LEGGATE M.D.

He started writing the article. He referred to his notebooks, copied out sections in which he had summarized everything known about the disease. He copied conclusions about cholera being a disease of the alimentary tract, and he wrote his new idea about the flies. Then he remembered something about flies in Sunderland … all those years ago.

He found his diary from that time. He had recorded everything. Autumn had been warm, with the harvest bountiful. A doctor to whom he had spoken had observed "immense flocks of young green toads."

He turned the page, and read what the same doctor had recounted:

. . . more than usual numbers of bumble-bees during the autumn, and dreadful plagues of flies, huge numbers of them.

This is the clue, he thought. Hardly anyone would know of this. Others think of miasma, of the atmosphere. Let them think—I shall consider the flies.

He went to lie on his couch. The voice of Magendie came to his mind: "You've beaten me to it."

It'll be a long article, Leggate thought. But leave the maps for now—just do the effects of the poison and the flies. I'll show it to Halliwell, and to Newcombe, to get their suggestions. Chemical analyses of poison, I don't have a laboratory, but Newcombe's a microscopist, could we visualize the poison? as particles were visualized in the air by a trick of sunlight? Perhaps we could see, under the microscope, the droplets of poison adhering to the flies' legs?

"That's right," he said aloud.

This could be it, really, he thought. There was always the impediment with miasmata that the atmosphere is far too general for such a specific influence. But this could be it. First the poison on flies' feet, then the flies on food, hence the poison reaches the gut. There! It's the flies, the connection no one has seen, the connection too between the maps, the filth, the stinks, and the alimentary mechanism of the disease.

Then long into the night he wrote of his alimentary theory, and of flies.

Chapter Seventeen

Early next day Leggate, eager to see his friend, had his horse brought round. If he could catch Halliwell before he went off to his factory, so much the better. Otherwise he would leave a note, with his article.

Halliwell was still at breakfast. "I think I have it. The connection between the immediate cause of cholera and its means of propagation."

"Well—tell me," said Halliwell. "Don't keep me in suspense."

"It is all in here. An article I've written for *The Medical Gazette*. I don't want to tell you, I mean not like this. I want you to read it, to see whether the argument's complete, or whether I should say more . . . or, more forcefully."

"Today's not so easily possible. I must meet some shipbuilders in an hour. Then with the harbour master about the design of sluices for the lock to New Dock. I'm taken up all day."

"This evening?"

"Well, no, again I am spoken for. Margaret and I are going out, but I could read it tomorrow morning, before breakfast. Why don't you come round tomorrow evening?"

"I'm being impolite. I can hardly wait to hear what you think."

He could scarcely contain himself for thirty-six hours. He rode towards home, a roundabout route alongside the great estuary. A train clanked in from Leeds. Leggate dismounted from his horse, looked towards the farther shore two miles distant, tied the horse to a scrubby tree. He walked to the high-tide mark, gazed across expanses of grey mud. The wind was a brisk south-westerly blowing clear down the estuary, the visibility no better than moderate. A newish looking ship, with its sails pulling, with a column of smoke extending forwards and its paddle-wheels churning mightily, Full Ahead, was rushing down the estuary on the last of the ebb, heading towards the North Sea, and the Continent.

Leggate, in a mood of agitation, thought: my idea crossing to France and Germany. It'll be in time for the next wave of disease . . . But think what's to be done, he told himself. Implications for preventative measures. I've not thought of that, yet that'll be its vindication. Any idea of aetiology must lend itself to prevention, or it is null. First . . . let's see . . . first, people must not expose their food to flies. That's it. Everyone must keep their food covered. And the flies, how do they acquire the poison of the disease? By walking in the heaps of dung, filthy things, and then on food, so nuisances must be removed; that's what's proposed by Chadwick, and by the Sanitary Committee here. So they're right, though no one takes removal seriously, but if the real reason why filth is deleterious were understood, everyone would rush to have them cleared. I shall write a second article: "Part II. Prevention of Cholera Morbus."

Leggate had walked some way eastwards along the strand. "Cover your food, remove nuisances"—that's not much of an article, he thought. Must observe flies, to understand their habits, so that prevention can be arranged more exactly. That will take a little while, but Part II can be published a few weeks after Part I.

Leggate came to a halt, and gazed into the grey distance, towards the Continent. This is truly it, he thought. At last it all makes sense, and I'll be seen as a benefactor. Lord Andervilliers will say, "Yes, my boy, it's a fine stroke of insight on your part," and Miss Brooks, she'll think . . . in that look, at her aunt's, it was as if she sensed something in me.

Frustration at waiting for Halliwell's opinion was not all that agitated Leggate. The visit behind the scenes in the Theatre Royal and the encounter with Miss Brooks at her aunt's bedside had intruded.

Though his last thoughts before sleep the previous night were of the article he had finished, he dreamed not of flies, not of a ribbon being pinned on him by the Queen to express the gratitude of a nation. Instead, as dawn broke, he woke from a dream of backstage regions at the theatre, and as he had lain in the warmth of his bed its influence had lingered in that half-world between sleep and waking.

Now, beside the estuary, the dream remained vivid, disturbing: a dream of searching backstage along deserted corridors, searching. I grope among eerie pieces of scenery, stage properties with distorted shapes, as if here behind the curtain the materials of life are being prepared for future events in which they will play their part, or have been already discarded so that they've become meaningless. My limbs are paralysed, but I'm searching for Miss Brooks, and with the paralysis it's a matter of utmost urgency, and then I glimpse her along a corridor, but I cannot move to reach her, and then Miss Brooks is Evette.

The dream's mood was difficult to dispel, a profound despair in his search for her… an utmost urgency together with an utter incapacity to move. Now, still, in broad daylight, as he walked alongside the estuary the discouragement of the dream lingered.

The visibility seemed to have deteriorated, the ship was no longer in sight.

Miss Brooks, he thought. Am I now dreaming of her! Have thoughts of her set Evette free?

On waking from the dream at dawn, when it was too early to go round to his friend Halliwell, Leggate had found his diary of September 1831, when he had first met Evette. He had not read these passages since they were written. All was here, recorded. It seemed essential, in the aftermath of the dream, to do something active, to counteract that paralysis, to read it and to take mental hold of those events, and master them. He'd laid the diary open on the table, and brought his two candles closer.

Leggate was a meticulous recorder of everything. If there were a distinct aspect of character, it was this obsession with observing and recording, with little distinction between thoughts on science and thoughts on the most private of subjects. They were all written in his graceful hand, sometimes a scientific idea jostled a personal thought in the same sentence—happenings, conversations, intimate actions, reflections.

It was in Paris, after having watched Evette in the same play for several evenings through opera glasses from the back of the gods, that Leggate had first ventured to those regions beyond the curtain. He brought with him a bunch of wild violets.

As dawn began to offer its first faint rays, in the Middlethorpe of 1847, he read what he had written in his diary those years before.

At the stage door, the doorman, whose job it was to keep out people like me, said *"Entrez."* The company was holding a party, and I was mistaken for one of the theatre folk, so before I knew it, I was on the other side of the curtain, on the stage, face to face with Evette Artaud. From close to, she was not beautiful in the usual sense, but more strongly than I had seen from my seat in the theatre, she exuded vitality.

"I have brought you these flowers, I've greatly admired your performances," I said. What audacity had brought me here?

"Why, thank you. Are you an actor, an author?"

"Just a medical student, assistant to Professeur Magendie at the Collège de France, and at the Hôtel-Dieu."

"Then how did you get in? Never mind. Why don't you join us. What did you say your name was?"

"John Leggate."

"I shall introduce you to the director." She led me to a thin-faced man, who was just filling his glass. "M. le Directeur, this is John Leggate, a brilliant young doctor at the Collège de France." She smiled in a conspiratorial way, squeezed my arm, and departed.

"Are you a friend of Evette?" the director asked.

"I just came in to give some flowers to her. I had admired her performance."

The director laughed. "A gate crasher! In that case you must pay your entrance fee, you must give me a consultation. I've got a pain, quite severe, indeed I scarcely think of anything else. Give me your opinion. My doctor is most unsatisfactory, he says it's because I keep irregular hours, and recommends quiet evenings, and early bedtimes. What does he think I am, a schoolgirl? Why don't you examine me?"

I was led behind the scenes to a room littered with books. Sketches and playbills covered the walls, there was a table piled with papers. I had scarcely met the man, but he spoke about the most personal of matters. "Here," he said, "I'll lie down." He stripped off his shirt, and lay half naked on a couch.

"Tell me about the pain, where is it?" I said. "Can you point to it? Is it sharp, or dull? Intermittent or continuous? When does it happen? Tell me about its severity and duration. Is it better or worse after a meal?" I pressed my hand gently on the director's abdomen, feeling for the spleen below the costal margin, and for the colon to see if I could detect constipation, then I pressed deeply and asked if this hurt, or that. I examined the director's hands, his eyes, felt his pulse. I asked about evacuations, micturition, dietary habits, the amount of wine he drank, and his patterns of sleep.

"Of course what I am able to do here, without any instruments, is quite limited. It may be that wine has been irritating you. This occurs in some people, but physically I can find no abnormalities."

"No abnormalities."

"Mlle Artaud was not correct when she said I was a doctor, I do work at the Collège, and at the Hôtel-Dieu, but I'm only a student."

"I can tell when a man knows his job."

The director seemed taken by me, and asked what I would suggest.

"I suggest the application of science. First we must make observations, so you could record on paper every time the pain comes on, or stops. Note when and what you eat, what you drink, and what happened just beforehand. Perhaps you could make a table, and record the condition every hour; it may seem tedious, but I believe it would benefit the case. Then I could look at the pattern."

"First sensible idea I have heard from a doctor in a year. Science, that's the ticket. Not mere opinion, we need exact knowledge. I shall do it. Come again next week, and we will get together after the performance. I'll bring you my observations. Now, let's go back. We're missing the party."

The director led the way through corridors, and past pieces of scenery. A labyrinth, stage properties, corridors; there was more space back here than for the audience.

"Come this way, I'll introduce you to the doorman. Perhaps you already know him."

"I was surprised he let me in."

"So am I. But on this occasion I will not reprimand him."

Winding through passages, past dressing rooms . . .

"Dr Leggate, this is M. Dujardin, our gatekeeper. Dr Leggate is a friend. Please let him in when he wishes."

"Yes, sir. Certainly, sir."

The director led me back to the party, and to Evette Artaud. "A good find you've made," he said. "I've appointed him resident physician."

He left the two of us together.

"I am sorry," she said. "He's the most terrible hypochondriac. You deserved it, but I didn't expect him to spirit you away for an hour." She looked at me. "Here, I will make recompense, I'll bring you a glass of wine, and introduce you to some delicious young women."

Though I don't drink usually, I took the wine, and she took me by the arm, making me feel an intimate. She led me to a small knot of people.

"You've made a diagnosis of our director?" asked one of them.

"The affliction of being surrounded by far too many charming women," I said. "The prognosis is dire!"

The young English medical student successfully treated the director by having him drink tea instead of coffee, by increasing the amount of fresh water he drank, and by making some other small modifications to his diet. Leggate was caught up in the camaraderie of the theatre company. He was taken by the affectionate way they touched each other when they conversed, and excited by the way his ideas were received—on literature as well as medicine. He walked through Paris, thinking of Evette Artaud. At the end of each week he would attend the party.

"I don't know how we managed without a resident doctor," Evette said one evening, as the two of us sat together, apart a little from the general company.

"I don't know how I managed without you," I replied.

We talked for a long time. I had a sensation never before experienced. Everything extraneous disappeared; it was as if something strong but indefinable were happening between us, like a feeling of awe when listening to a profound piece of music, but not individual, it was between us, palpable in a way we both knew and understood.

"Let's go," she said. We walked through darkened streets, scarcely talking, her arm linked through mine.

"Here we are," she announced, producing a key that opened a small door set into the great door through which carriages pass to and from the street. We ducked through, walked to the opposite side of a courtyard, up curving stone stairs. There opened a small apartment, with a drawing room, furnished as if in a stage set.

"You would like to stay with me tonight?"

We were a foot apart—it seemed the most obvious thing in the world to draw her to me, to kiss, her softness of lips expressing an infinity.

"You sit here," she said after some minutes of this. "Would you like a drink? Some brandy? I won't be long."

"Nothing, thank you."

I sat on her *chaise longue*, with my eyes drifting from object to object. Is this to be what I have longed for all this time, I thought.

She returned, wearing just a shift. "Come." She extended a hand. I followed into her boudoir.

"Let me help you with your cravat," she said, standing close, looking upwards slightly, directly into my eyes.

"I haven't . . . I mean, this is the first . . ."

"Ahhh, the doctor is a virgin. Put your clothes on the chair there, then get under the bedclothes. I'll be back in a moment."

She left the room. I removed my clothes though not, of course, my shirt, and climbed into the bed. Peeping over the covers I saw her return, carrying a candle, which danced as she walked. She had extinguished all other lights. She placed the candle on a table beside the bed. She crossed her arms just below her waist to grasp the material of her shift on each side, and drew it quite slowly upwards, as if the curtain rose in the theatre. For weeks I had been in a fever, anticipating, but not knowing how to anticipate this moment. Now it was here. A simple gesture, starting with the hands low down, rising steadily with continuous movement to end with the garment stretched now above her head, was the opening to a world for which I had longed with all my being, but had been unable concretely to imagine. For a long moment she stood, then dropped the shift on the floor. Then she tossed her head gently, lifted the bedclothes and slid under them.

"Here," she said, "come over here." Then: "But this won't do. Take this off." She threw back the covers. I sat up. I pulled my shirt over my head, and pulled up the covers again.

"This is the first lesson," she said. "No shirts. You must feel this touch of skin on skin. It makes a special kind of sensation that cannot be made in any other way."

I did feel it, her skin touching mine. The strangeness of another being, closer here than I ever imagined. Her ankle curled round one of mine. A hand at the nape of my neck drew my head towards her own. Parts of the body for which I have no counterpart were pressed to my chest, and a tender softness enveloped me as, with the hand still on my neck, we kissed. Feeling perturbation lower down, I drew back a little—only to have the hand leave my neck, travel to the small of my back, draw me towards her more completely.

"There," she said.

I could not reply. I found her mouth. I was encircled, borne up, complete. As if some lost aspect so long absent were now regained. A meeting of a secret kind with another complete creature, like me, but so different from me.

"But, we should not," I said. "I mean—it would put you in danger of ..."

"Of conceiving? It's not to worry. I cannot. When I was younger, I had to stop a pregnancy. I got an infection and nearly died. I was told then I could not conceive." Her voice was sad.

"I am sorry ..."

"Perhaps it's for the best. Now," she said. "You come here, you."

She lay on her back, took my hand, placed it on her belly and guided it in tiny circles, quite softly, quite slowly. Guidance did not have to be protracted. I began an exploration. Above, below, behind, a world enough, and time. Then, abandoning a certain coyness, I knew not how, unless it was because of her intimate touch, I threw back the bedclothes, turned and rose on my knees. An hundred years should go to praise. She smiled tenderly. In that moment my thudding heart calmed. She reached her hand towards me, changed the position of her body a little, and drew me to her. Then gently down. She pulled up the bedclothes over my back, and placed her hand again behind my neck.

So began a period of Leggate's life when he was no longer an earthling. They grew together, as if a vital connection, once utterly lost, was remade.

"Do you feel it?" she asked. "The sense that we are part of each other."
"As if we were the very first people."

To her he brought devotion of which he had a great store—exhilarating to an actress—and a youthful intensity. To him she gave a warmth, a tenderness, almost motherly. He would write about his transformed state:

It is as if I walk inches above the street, and hours float rather than drag. I feel a benevolence to all. In the Jardin des Plantes, at ten in the morning we hold hands like children. A tree transforms so that every one of its thousand leaves is outlined perfectly, and every tree reveals its very treeness.

He wrote a small poem to give to her.

A sailor now has landed on this shore,
To find the place where lilies grow,
An opening, an unsuspected door,
Admitting one who thought, but did not know.

Together they discovered an attachment like a cord that joined them, on which either could give a small tug that would be reassuringly answered, as if it had been from the beginning.

Now Leggate, on the shore of the muddy estuary in Middlethorpe, realized he had walked nearly half a mile, and had almost reached the entrance to the Albert Dock. He turned to face the wind.

The idea of a chain of causes of cholera, he thought, and now ... Miss Brooks, and I need to concentrate, but I can't until I hear from Halliwell and what he thinks about my article, and now thinking about this. Quite unlike my situation in Paris. Taking a wife is possible now ... Is she already promised? Was it accident she was visiting her aunt? Her look meant something.

I didn't imagine it. Perhaps she's been waiting before marrying, as I have waited, not just a general practitioner, soon a scientist, conducting correspondence with others in Paris or Padua. If this discovery about flies is right, I'll be elected to the Royal Society. Before now it would not . . . But soon, a man of whom a woman would be proud.

He started walking back towards his horse. Why tell her aunt she was going home to Sutcliffe Street? he thought. Her aunt must know where she lives. She wasn't telling her aunt. She was telling me.

"That's right," he said out loud.

Leggate knew the town as intimately as any native: Sutcliffe Street was a short street with houses of substance. He walked back quickly to untie his horse. When do ladies go visiting? Four o'clock? See my patients, and get everything done, then I'll walk that way, on the off chance. If I happen to see her, it means something.

A little after half past three, Leggate shouted over his shoulder, "I'll be back in a while, Mrs Huggins." He took his hat, slammed the front door. Dodging between amblers and women carrying parcels, around pedlars and tradesmen, he ran across the street in front of an overloaded cart. He was sworn at, he splashed by mistake in the mud, then found himself after fifteen minutes walking in a more measured way along Sutcliffe Street, one end to the other. He recognized he had been here before.

Which is the house? he thought. Fancy not remembering, but her father is the rector, the house must be near the church, I suppose, but which one?

At the end of the street, he looks down a side street to the left, walks down it a few yards. A carriage rattles out from a mews, comes towards him, passes. He looks round to see it turn into Sutcliffe Street. He walks back to the corner and looks after the carriage. It halts a hundred yards along the street, on this side. Leggate retires behind the corner. After an interval in which he feels distinctly his heart pounding, he sets off, he rounds the corner, now fifty yards, now first Mrs Brooks, and then Miss Brooks, descend the front steps of a house. The carriage door is held open, Mrs Brooks is handed up. Ten yards. Five yards . . . now two.

Miss Brooks glances in his direction. "Dr Leggate! I wouldn't have expected to see you here."

Mrs Brooks's head emerges from the carriage doorway.

Leggate raises his hat. "Mrs Brooks, Miss Brooks. How do you do. I was taking a short cut. From a patient . . . in Green Street."

"May we take you on somewhere?" asks Miss Brooks.

"No, thank you, really. I expect you are going visiting. I bid you both good afternoon."

"Good afternoon to you," says Mrs Brooks, withdrawing her head.

"Good afternoon," says Miss Brooks. Then her face is suddenly close, whispering: "An uncommonly interesting patient you are visiting without your medical bag, doctor."

She seats herself, pulls down the window, smiles to take the breath away. He raises his hat once more. With a shake of the reins the coachman sets the horses forward.

Leggate stood regarding the back of a carriage as it clattered up the street. It was not science, but for him, in this meeting, his conjecture of her interest was confirmed beyond all possibility of rebuttal.

Chapter Eighteen

Impatient to hear his friend's opinion of his paper, Leggate was next evening at Halliwell's immediately after dinner.

"It's a fine leap of inference, the ring of truth," said Halliwell.

"You mean the idea that flies spread the disease?" said Leggate.

"The idea that the disease must first be due to some destruction within the mucous membrane of the gut, so blood vessels there become permeable, and therefore that a poison must have been swallowed, and must be the effective agent."

"A distinct poison . . ."

"That's the boldness of the idea. A poison that one can actually extract, or distil."

"You think the argument compelling?"

"It has a rightness about it. It is specific . . . if a pump has gone wrong, it's because a part has cracked, not just because it's old."

"But the mode of transmission, the flies?"

"There I am not sure."

"You mean the argument's not compelling?"

"It is an idea—there is no argument as yet, is there?"

Leggate was acutely sensitive to criticism. There were few from whom he would have tolerated it. Halliwell was so good-natured that Leggate could expose ideas to him, even at an early stage, without fear of being wounded, but now his friend's words cut him.

"I see . . ." he said.

"An editor might not find it compelling."

"I need more evidence?" Leggate's face fell in such a way that he knew his friend would realize he had gone too far.

"It needs intermediate steps . . ."

"Between the idea of poison, and the idea that flies deposit it on food," said Leggate.

"Why don't you send just the first part to start with? Argue that the disease is an affliction of the gut, probably by means of a poison taken by mouth?"

"That's not much forward from fifteen years ago. It's advanced somewhat . . . the substance taken by mouth. But it presents a problem: those who believe in miasmata think it's in the air, that you can smell it. Without something new they would resist the idea that it was swallowed."

"How does the poison get onto the filth in the first place? There's filth all the time, but now we don't have cholera."

"I can't see it without the flies. Flies could come bit by bit. They could explain why the disease travels so slowly over land, a mile or two a day, no farther than they could fly, and explain how the disease makes sea crossings, because the flies could live in the hold of a ship, but you say the flies are not compelling."

"The poison must be able to act in tiny amounts."

"My solution was an army of flies depositing it on filth where it stays until other flies transport it onto food, but I can't altogether see how it could work in such minute amounts."

"Hmm."

"Though the poison's in tiny amounts, its effects are in the large. In Paris people died by the thousands."

Halliwell has spotted the weakness, he thought. It's there for me to solve . . . but out of reach.

"The poor are susceptible because they eat meat that's old," Leggate said, "after the best cuts are sold. Have you seen meat that poor people consume? Walked on by so many flies that it stinks."

"You could study the ways of flies."

"But you think I am not yet ready to publish?"

"Begin by publishing the first part, then see how that's received. If it's received well, you can go ahead with the flies. But the flies are ideas, are they not, at least for now?"

As when a piece of iron cast imperfectly is set aside to be melted down again, Leggate felt he must begin anew.

Is the poison here all the time? he thought. If so, could the disease depend only on the number of flies that carry it to food? . . . so that severe epidemics occur when flies become numerous.

He visited two patients, wrote nothing in his notebooks, lay on his couch, stared at the ceiling. In the afternoon he crossed North Bridge, walked south to the shore of the estuary, then walked eastwards along the high-tide line.

I'm not cut out for this, he thought. I felt sure I was right, then Halliwell, not even a doctor, saw holes. I'm wasting my time—all those maps.

He picked his way over gravelly foreshore to the water's edge. Looking down he saw he had disturbed a small crab. The tide line teemed with life that was not obvious until one looked with care. He stooped to peer, as the crab buried itself in the mud.

As well I showed my paper to Halliwell before sending it off, he thought. He emitted a loud sigh, unheard by anyone but himself. I would have had to write and ask for it to be returned. Mortifying. I could never then have sent anything further to *The Gazette*.

Leggate stood up. A seagull gliding close uttered a cry, then soared into the air . . . Peerless grace . . . Miss Brooks, he thought. What she said as she got into her carriage had a knowingness in it. She'd want me to find the solution.

Next day Leggate woke before five, energy renewed.

Definitely good to have exposed it to Halliwell, he thought, as he dressed. The first part's correct! I must take one step at a time. So that's an advance. It's premature to publish the flies yet, I must study their habits, and then have the whole train of causation.

It was early. He read again through the first part of his paper.

Halliwell said this part about the gut is correct, he thought. So we've narrowed the problem. What is the problem now, exactly?

He started writing:

How can enough poison be transmitted? Poor people eat food that has been left out so hundreds of flies have been on it, that would explain the prevalence of diseases in the summer, not just cholera but other diseases with milder forms of diarrhoea.

So this is an advance. Not yet the critical one that will convince doubters —if this were such an easy matter, others would have reached the conclusion, but the field is narrowed.

That evening, after he had eaten the dinner his housekeeper had made, Leggate's thoughts strayed to regions from which they had long been excluded. Once more he could contemplate Evette. The years in Paris took on another meaning: a prophecy of the possible. We couldn't marry then, he thought, but now I have means and prospects, both.

Leggate felt again on the edge of that region of grace where earthly cares are transitory, where each leaf shines as the light strikes it. He could take a wife. What better than a wife who would fill a house with music. Reward for endeavours, reason for existence.

The path of my life and of Evette's only crossed, he thought. They couldn't run beside each other, because for her to leave the theatre . . . for what? for me to be an apothecary? That was intimation, but now—the look Miss Brooks gave me at her aunt's bedside, then from the carriage, they were acknowledgements.

His reveries of the success of his researches had been strong today, but thoughts of this other kind were stronger. Days of science would be adorned with evenings of music and affection. Would she not want to marry?

What was it, all those years ago, he thought, that I tried to capture in my diary? I must find the place, I haven't read it since it was written. It wasn't for reading then, more for just crystallizing it, and later too distressing, but now? . . . Now a guide to the present?

He took up his diary labelled "1834," turned pages, came upon this passage.

Evette's play was interrupted for four days by some other piece that she was not in, and at the same time I got permission to be away. We left Paris, and went by stage to where Evette's friend Madeleine lived.

"She did what many girls dream of," Evette said. "She escaped. A wealthy man fell in love with her. Now here she is, married, with children whom she loves, living in a great house."

It is not such a great house. It is a small château, but beautiful. At breakfast one helps oneself from the sideboard. The first morning, Raymonde was already out.

"Making improvements to the estate," Madeleine said. "It must be the most improved estate in France."

The two women conversed while I looked through glass doors at the garden, across a green lawn with a ha-ha to stop the deer coming up to the house. Then the view extended across a field that sloped towards a lake, which seemed to have been especially placed to please the eye.

"As if we are standing at the edge of a painting," I said.

"I shall tell my husband you said so," said Madeleine. "For twenty years he's tried to create such effects. Before that his father did the same. He was not a grand aristocrat, and fortunate in the Revolution. 'Painting with nature,' my husband calls it. He says he doesn't try to outdo nature, like some gardeners, but to act as its assistant."

Madeleine came over to the window. "There's a small river that runs in the valley there, see?" She pointed. "My husband's father had it dammed to create the lake and this view from the terrace with the little stone bridge as the river enters the lake. He planted trees, brought as seeds from all over the world. Why don't you and Evette go for a stroll. It's quite beautiful."

"Shall we?" I asked.

"I'd like nothing better," Evette replied.

"Over to the right," Madeleine was pointing again, "you can see a boat-house, on the other side of the lake. You could go for a row. Can you row?"

"A little," I replied.

"You'll come to no harm. In the boathouse, take the smaller of the boats. You'll see some cushions there that fit it."

When we had finished breakfast, we went onto the terrace. The path led down, across the little bridge, then through a grove of trees. After a fork, it swung to the right towards the boathouse. Evette wore a simple dress with a loose skirt that reached to her ankles, and with a bodice that left her neck and arms open to the air. The yellow straw hat Madeleine lent her offset her black hair which she had let down. Madeleine lent her, too, a rose-coloured parasol, to shade her arms from a sun that stood stationary in an azure sky.

If one should want to know Eden, this would be it. After the grove we came out onto the path across the field, where a small herd of Guernseys stood grazing almost as if they too had been placed to enhance the sight.

"When Madeleine grew up, she didn't know such places existed. Nor did I," said Evette. "If Raymonde had been a grand noble, he wouldn't have married her. But he had this house, and money from some concern in Paris. He has no family to please, so he pleased himself—with Madeleine. Probably it's a scandal, but they are happy, don't they seem so to you?"

"So it is just a matter of making the right catch," I replied.

"Don't sneer," replied Evette. "I should scarcely have taken up with a penniless medical student."

"I'm corrected."

"You men have not the faintest notion. You and I have been without conflict because I have not opposed you. But women have very few choices. If you could see our paths from the beginning of our lives to the end. Here . . . paths like this one, across this field."

Evette kicked at a clump of grass. "Just three or four paths, each with little distensions, one for each significant event. The most trodden is marriage—a man, then the wedding, then a child, another child, a child's death, then another child, until the body is used up, then death.

"Then there's a smaller track that runs parallel, a man, no marriage, no children, loneliness, death. Then there's another path, much disapproved, first a man, then another man, and another, and so forth, then death.

"One can stray from any of them, even occasionally leave one and join another, though it's never advised. To make a living there aren't many possibilities. Men, on the other hand—well, men of a certain class—are kings of infinite space. You can journey across the field in any direction, needn't follow a path at all. You can visit the Indies or the Americas, you can write plays, govern cities, make great discoveries like your M. Magendie whom you adore . . ."

I refused to be drawn. "Which path would you follow?" I asked.

Her eyes flashed, unnerving me. "I won't take these womanish paths."

On the other side of the field, there was a stile, which we climbed. The path wound through a small clump of trees, then out into the open again. There was a bench, with yet another view across an uncut meadow towards the white painted boathouse on a little promontory. She sat on the bench, looking at me.

I was shocked at her vehemence. She must have seen my surprise and said, "You're a baby, John, a child—and you think I'm your mother. With your cleverness and your studies of people in health and sickness, you don't know how the world works, or for whom it works."

I sat down, and gazed forwards, seeing neither trees nor lake. I found it hard to bear her anger, and could not understand what she was saying.

"If you had money," Evette continued, "I might leap into your arms, for I believe I love you. But as it is, we would get immediately onto one of these paths. If I weren't to work, you would be weighed down with debts. I have my mother to look after. You've not finished your training, you'd never make the discovery you crave, and I . . ."

"We could stay in Paris, you could continue on the stage."

"It doesn't work like that. My play will come to an end soon. I have nothing else."

"We would manage."

"I don't believe we would." Evette folded her parasol, laid it on the seat beside me and rose. I rose to go with her.

"Stay here and look at the view, I'm going to pick some flowers."

I saw her walking away from me, in her straw hat, brushing her way through long grass, her skirt swaying, the world contracted into a span. In the picture of this garden, across the lake with the house now small in the middle distance, the human form had been lacking. Now her angry words of impossibility transformed it into another world, unreachable.

A wave of loneliness came upon me. Her denunciation had made a gulf, left me on this other side. I had asked her to marry me several times. Always the answer was the same. I was impractical.

I watched as she bent to pick flowers, first here, then there. An evening came to mind, the next after the first night we had spent together, when she had told me about how she became an actress.

Her parents had run a tavern near one of the theatres. Downstairs, people could meet, eat, and of course, drink. Upstairs, there were rooms for travellers for a night, or a week or, for those not travelling, for an hour or an evening. There were younger brothers and sisters. As they became old enough, the children had worked at the tavern. She told how in her childhood she had received as many blows as other children, precociously she had witnessed things that aristocratic ladies never dream of. As a child her mother had protected her to some extent. At fifteen she had been noticed by a theatre director.

"I was a comely wench," she smiled at this thought.

"At fifteen, what degradation."

"*Au contraire*. It was my salvation. From him I received more kindness than I had known. He taught me acting, the theatre, literature. He was my university. When I was nineteen, he said, 'We must part. You're old enough to go your own way.' He had, of course, a new protégée. I was beside myself with jealousy. He gave me a talking-to. 'I've given you four years. Do you think you were the first? What about the others who came before you, I set them on their feet, do you think they were not jealous? You can say I'm treacherous, a scoundrel, but I've not made you pregnant, I didn't promise marriage. I couldn't live in that way. I always said this was just for a time.'

"Of course I'd taken no notice. He gave me money. 'You should get your own place,' he said. I see him sometimes still. He's a charming man, still kind. After a period of despair, I picked myself up. I realized I could earn my own living. I keep myself."

"And when there's no play?" I had asked.

"I turn to other resources."

I was roused from my reverie, by a swishing through the grass. She walked towards me, with the sun behind her so that I could not discern her features, merely a too-bright shadow, and a bunch of wild flowers held towards me.

"For you." I took the flowers. In an instant she was running towards the boathouse.

"Can't catch me," she called.

I rose, with the flowers now in my hand, about to run, when I remembered the parasol, picked it up, then ran in pursuit. She reached the boathouse before me, and turned her face as I arrived, laughingly gave me a kiss on the lips.

"I said you wouldn't catch me."

We entered the boathouse: a large wooden room with decking on each side, and water down the middle leading to an archway opening onto the lake. On a bench were two boat cushions, each four inches thick and three feet square, for the smaller boat. I fitted one on the stern seat, and stood the other upright against the wrought-iron backrest.

"The rower doesn't need cushions," she announced. "He must exert himself."

I found the oars, held the boat while Evette got in and seated herself in the stern. I remembered her parasol, and passed it to her.

As I write this, I remember the sad little thought of that moment, that the flowers we left on the bench would wither. I untied the boat, and got in myself, and pushed us through the arch. I put the oars into the rowlocks. The scrunch of each stroke alternated with the sound of water rippling beneath our bows.

"The sun," Evette said, leaning backwards in her seat, her eyes closed, her face upward. Her hands held her straw hat.

"You should use the parasol."

"I like the sun. But perhaps you're right." She lifted the parasol, holding it to frame her face. "Row to the other side of the world."

I rowed more slowly, dipping the oars gently into small noiseless swirls. A bee buzzed past in search of some summer flower, circled for a moment round the boat, and then resumed its journey.

"Mmm," she said, sitting up. "Alone in the midst of the ocean sea."

She now leaned towards me. Her dress with a wide round neck had four buttons, now I noticed undone. I could not help but see her form, not completely revealed but in such a state as to suggest unseen parts that I could imagine. I looked away, but then could not restrain myself from glancing again. I looked behind me to see where we were headed, then towards the shore, since it is not right to stare. Why is this sight so irresistible? I could not keep my eyes away. I looked up at last to meet her eyes and found these eyes regarding me, smiling.

"At last," she said. "I was wondering if I had become old and ugly, or would have to become a mermaid."

Now I smiled too, blushing crimson like a boy.

She put down her parasol. The boat was perhaps four feet wide, with five or six feet or so between the stern seat and the thwart on which I sat. She made the boat lurch as she placed the cushion on which she had been sitting on the floorboards in front of me, and slightly under my seat, then sat on it, between my knees, nuzzling her side against my lap, so that I could no longer row.

"Mmm," she said, once more. She took my right hand from the oar, placed it inside the bodice of her dress. We drifted in the middle of the lake, with only the faintest breeze.

"Perfection," I said. "Let's stay for a million years."

"Before we are shipwrecked, why don't you row over to where those willows are." Evette arranged the second cushion also on the floorboards so that she could lie full length on her back on the bottom of the boat with a bare foot in my lap, where she was easily able to ensure the continuance of a certain physiological state.

I steered the boat beneath overhanging willow leaves. Evette rose, took an oar from its rowlock. She thrust it into the mud, to hold the boat securely against the jutting roots of the trees.

"I have seen boating people do that."

She kneeled on the cushions, spread like a small bed. "Now you lie here."

A second later she was undoing my clothes, then quite quickly she knelt with one knee on each side of me, on my lap, the skirt of her dress decorously spread to cover us, her face above mine as she leaned down so that we could kiss.

She lifted her face. "Now we will see that there is also another state of perfection," said she. I glanced again down the bodice of her dress, closer than before. We kissed again, perhaps for a century. Now she lifted her face once more, and sat upright. Pulling down first one of the short puffed sleeves over her elbow, and then the other, then bare shouldered, bare armed, not neglecting to regard my rapt concentration, she lowered the bodice of her dress to her waist.

"I want," she said, "a medical bulletin. I am over thirty years of age, doctor.

Do you notice wrinkles? Unseemly pendulancy? Signs of disease?" She cast her eyes downward demurely.

"I believe you're in perfect health."

"That," she said, "is because I am not mortal. I come just for a short time to earth to take my pleasures here. I have chosen you."

Slivers of sunlight slanted through the willow leaves. Evette looked down at me. She raised herself on her knees, then with a small movement of her hips, lowered herself again, with the utmost delicacy.

"Now," she said, "concentrate fully, and fill my mind as well as my body with you. Just place your hands here." She placed her own hands on my shoulders.

She studied my face, then closed her eyes. Then with movements that neither of us made individually, and because as movements they were almost imperceptible, we reached that state in which not just our bodies but our entire beings joined. Something like a blush started at her neck, and coursed across her upper body, till our souls then, I believe, truly did meet. She let out a little cry, then lowered her head to cradle it upon my breast, and lay upon me like an infant.

For two beings to come so close to each other in this way; is this the natural way of things, or vanishingly rare? How could one know?

She raised herself a little, looked at me, and said, "But not yet you? You are uncomfortable? No?"

"Quite the contrary."

Now again she raised herself, placed her hands on either side of my shoulders. We began slowly, tenderly, with certain movements of the hips— and all the while, this vision. I felt myself begin to tremble, uncontrollably.

A thought from nowhere: Is this what an infant sees as his mother leans over him, when she is everything in the world?

At the same moment when I reached that state, she underwent another agitation. She pushed herself downwards, grasping my shoulders, pulling me towards her. I held her head between my hands, and uttered a wish.

Is this the state in which the yearning of imagination is matched exactly by the presence of the other person? the connection at the heart of things? I held her head on my shoulder, my arms enclosing her, my eyes closed. My mind soared to some high point in the heavens, above the willow trees, from where I could see our little boat again in the middle of the lake as we drifted silently on a blue expanse.

"I must have dozed," she said. "Have I made you numb?"

"You've made me more contented than I thought possible."

Chapter Nineteen

Marian said to Caroline, "I shall see you're asked to the Larkins' dance. I've arranged for Dr Leggate to be invited as well."

"A plan," said Caroline. "So I shall meet him?"

"You and your husband. You can chaperone me, so my mother can be off duty. And I shall want your opinion."

"You'll give me a list of special points to look out for?" said Caroline.

"Don't be so cruel," said Marian. "You know I feel guilty about attracting his attention, but he's not been out much in company, so now he's coming to a dance—isn't that how one's supposed to meet someone?"

"I'm sorry," said Caroline. "Of course I want to meet him."

Marian turned to a matter that was agitating her mother. William had asked Mr Brooks for money. He had not chosen the time well.

"It was just after Father received the bill from the dressmaker," said Marian. She pursed her lips. "William asked for a sum—I don't know how much—to pay a gambling debt from billiards."

"Your father was upset."

"Furious. He understands dresses. It flatters him that my mother looks well, and she insists that I do too; well I try to at least. But money wasted in gambling!"

So in the Brookses' house there had been dissension. When Mr Brooks received an appraisal that he had ordered on a farm for sale ten miles the other side of Battersby, which he had been considering as a place for William, he was displeased with it.

"My mother has suggested that William goes to stay with her brother James, in Canada," continued Marian to her friend. "Apparently my father said: 'He should be transported to Australia.' Then my mother said fuming would not solve the problem, which must have made him even more cross.

"But a day later he asked her to write to Uncle James to see if he would take William for a while, to teach him the rudiments of agriculture."

Giving a passable imitation of her father, Marian said, "You put it to the boy. If I say he goes to Canada, he will think that is the last thing in the world that he would do."

"Would William go?"

"It would be good for him," Marian said. "He's become one of those people who exists to avoid things. He drifts about without plans of any kind. He ought to do something. He has no sense of direction."

Plans, thought Marian. What have I been doing since I came back from Leipzig? Until Andrei, I had a sense of direction, and since then just a senseless yearning, for something. What have I been doing except drift?

Leggate turned back over the pages on which he had written of himself and Evette in the garden, in the boat on the lake, under the willows. The words he had written seemed to detach themselves from the paper, to agitate each other.

The week after Leggate and Evette had returned from their visit to the small château, the director and the owner of the Ambigu had decided that the new piece which had been tried out would go on for a season, and after two further performances Evette's play closed. Between Leggate and Evette there had been uneasiness, irritability. Then had come the long letter which he, unbelieving, had read. Then he had reread it. She loved him, but a life together was impossible.

The letter had been the last he heard. Her rooms were empty, her furniture gone. He imagined that the director had reclaimed her. Or was this tit for tat? so that just as she had been given three or four years by her first lover, now he was being given the same. He imagined a wealthy manufacturer. He imagined her with an aristocratic owner of an estate with a lake. He even imagined, in a perverse scene, her joining her friend Madeleine so that Raymonde would enjoy both of them. More painfully yet, he imagined her on a street corner—a woman with despairing eyes making coarse suggestions.

Not until now, thirteen years later, had he been able to contemplate his loss. He closed the diary, consigning the words once again to their pages. He was now able to weep for the loss of her, and for the world that makes such things possible and impossible.

As his tears ceased, he thought, She was right to leave. It was ... what? A prefiguration of the possible? Now I have a practice, and the moment has come at last. Lines of poetry, I can't remember the rest ...

Who ere she be,
That not impossible she
That shall command my heart and me.

He walked to his bookcases, looked until dampened eyes alighted on a volume of poems by Richard Crashaw. He opened it, leafed through, and then came back to the table of contents. He found the title: "Wishes. To his (supposed) mistress," and turned to the page.

"Who ere she be ..." he read.

This is it, he thought. I'd forgotten. I got it second hand in the English bookshop in Paris. I haven't opened it since—haven't opened anything really, except science. Poetry, novels, all gone sour—only science. I've become some Faust, giving up everything.

He read the poem through ... "May she enjoy it, whose merit dare apply it." Then came an idea, unbidden: lend her the book, as if she'd requested it.

The binding was not worn. He found a slip of paper on which he wrote, "The book you wished to borrow," without a signature. It might mislead a servant or even a parent, but start intriguing thoughts in the recipient. He placed the paper to mark the poem.

He gathered his coat. A little while later he was at the Brookses' house. His intention was to leave the book with a servant. He knocked, and the door was opened. A servant said, "Come in, sir. Give me your things. Please go up to the drawing room."

Leggate did not know that Mrs Brooks received visitors on the first and third Thursdays of each month. He did not ask that the book be handed to Miss Brooks. He went up.

Mrs Brooks saw him, and said, "Come in, doctor. Good of you to come and see us."

To her husband, who was at that moment passing, she said, "Robert, you remember Dr Leggate? He was a judge in Marian's competition."

"Indeed, indeed," he replied. "How do you do." He continued his progress.

"I don't mean to interrupt. I didn't know," he said, looking at Mrs Brooks, "that you were receiving people this evening. I just wanted to drop something in."

"I'm glad you've come by," she said, with an engaging smile of the kind that only confident women bestow.

Leggate glanced around, saw Marian sitting at the other side of the room, in conversation with a young man.

"Dr Leggate—Mrs Steen." Mrs Brooks introduced him to the woman with whom she was conversing. "We were talking about this frightful murder. Mrs Steen tells me that in the paper this week the identity of the culprit will be revealed. It's quite spookish."

"My son knows the nephew of the newspaper proprietor," explained Mrs Steen. "He says that the news will be shocking."

"I'm glad our police station has a man to deal with it," said Mrs Brooks. Leggate saw that this was the kind of conversation in which he had merely to nod in acknowledgement. Mrs Brooks continued. "I know ugly things occur near the docks, but really! In the middle of town."

As the conversation ambled forward, Marian found reason to join her mother.

"Dr Leggate." They shook hands, a gesture into which, with the minute lingering on the occasion, it was not difficult for Leggate to read significance.

"I was on my way past, so I thought I'd drop off the book that you . . . ," he looked towards her mother, ". . . that Miss Brooks was interested in seeing."

Marian raised her eyebrows, a momentary expression. She took the book, opened it where the slip of paper protruded, read the message, glanced at the open page: "Wishes" she read aloud. She raised her eyes fully to meet his, smiled.

Leggate noticed Mrs Brooks looking at her daughter enquiringly. What a flimsy pretext, he thought. I only spoke to Miss Brooks alone for a minute or two after the competition, and now I make out that she asked for a book! But I don't mind.

"This is it," said Miss Brooks. "You're kind to remember. Would you like a cordial?"

"Thank you. I would like to, but I must be on my way."

Chapter Twenty

"Each thing in the body can be comprehended," said Leggate. "It works in a definite way, not a mishmash. Cholera can't be caused by vapours, also by squalor, also by immorality. There must be a single cause." He had struggled with how to write his paper. Now he was trying to explain his idea to Halliwell.

"A single cause?"

"Like a machine—you said so yourself. If a pump stops working, you find the cause. It's the same with the body. I know the immediate cause, fluid has leaked from the blood, so the next step is to find the effective cause, the poison that makes the vessels leak. How does it get there? By mouth—so by food. That's logical, but how does it reach the food?"

"You still think by flies?"

"I don't know. The amount of poison carried on flies' legs must be so small. I'm cracking my head."

"Don't give up, you must persevere . . . Here," he continued, "come and see my herb garden, it's good at this time of year." His hobby was botany. In summer here, near the fifty-fourth degree of latitude, evenings lingered. They walked through the fragrant herb garden—Halliwell picked a mint leaf, crushed it between thumb and forefinger, and gave it to his friend to smell. Then they climbed a stile, took a path round an unmown field of grass with wild flowers round its border.

"God is infinitely cleverer than us," said Halliwell. "See this dandelion." He picked one. "These little parasols." He pursed his lips, blew towards the sphere of feathery seeds. The two of them watched with infant fascination as the seeds floated across the field, buoyed by a tranquil air. "If one germinates, it grows to be parent to these things. Exquisitely fashioned, and their purpose? To make more of themselves. When we make something, we make just the thing itself. Our things need tending, need to be oiled, cleaned, repaired. If they break, you're right, we must find the cause—they don't last long. But these things renew themselves until eternity. It may take hundreds of years before we understand even a little bit about them. Each one who turns his mind to research makes only the smallest step . . ."

"I used to go and hear Balfour's lectures when I was in Edinburgh," said Leggate. "He was professor of medicine and botany. He would say that botany wasn't just for medicinal herbs, but for the deepest principles of life. He was interested in dispersal. Said a dandelion produces 240,000 grains of pollen."

"I've not read Balfour, but he's right. Dispersal's the key, each plant with its own means, for pollen and seed. Some use water, dandelions use the air, some plants use insects—like your flies. And within each fertilized seed, a tiny germ of the plant, designed exquisitely, ready to take root."

"Hard to think of machines ever doing this," said Leggate. "All the same, don't you believe these things, and the body, really are kinds of machines? They have properties of multiplying, but even so, they work according to causes we could understand." Then subdued, he said: "At least that's what I believe."

Halliwell went on as if he had not heard. "Can you imagine how God made each different species? They propagate themselves, and multiply. A dandelion now, really, 240,000 grains of pollen? Then all those tiny parasols."

Halliwell plucked another dandelion, blew gently, wanting to see again what he had seen hundreds of times. "They go everywhere, including my garden." He laughed. "They're exquisite, but where I don't want them I pull them out by their great long roots and burn them. Or there, these burrs on your trousers, you see?"

Halliwell took two burrs from Leggate's trouser leg. "If I dropped them, in a little time they would become bushes. Then multiply again. You were to be their means of carriage. I shall foil them." He crushed them with his thumbnail. "There."

They returned to the house. "You have something else on your mind, not just science."

"I have?"

"I would say so."

"Well, perhaps. I am feeling . . . let's talk when Margaret comes down."

The children in bed, Leggate, having started the wheel of curiosity, was impelled to speak.

"I am somewhat attracted to a young lady. I mean . . . I have only you to tell."

"You're in love! Who is it?" said Margaret.

"I didn't say I was in love."

"Come on, who?"

"Miss Brooks, daughter of the Rector of Swanscombe. She won the piano competition that I judged."

"And she . . . ?"

"She may . . . have some interest . . ."

"Oh, John, how wonderful. A musician. How fine that would be."

"You don't think marriage would be an impediment? Or take my mind too much from science? Some days I think I've almost reached my objective."

Chapter Twenty-one

Since the days when Marian had first attended dances to advertise her willingness to be invited, tempted, courted, these occasions had palled. The little orchestra, the tunes they played, had become banal, the occasions merely obligations that emphasized her state—unmarried, perhaps unmarriageable—so that the ceremonies of display and invitation aroused within her a satirical commentary that she found difficult to keep to herself. But at the Larkins' dance, she arrived soon after the start, not just with a renewed eagerness, but with a barely contained excitement she had not felt before.

A large room had been cleared of furniture, except for chairs around its edges and four grouped in one corner for the little orchestra. At the opposite end, a doorway led into a room where tables for cards were set out. Next to that was another room where refreshments were laid.

The quadrille was coming to an end when she saw Leggate arrive, look into the card room, then into the dance room. The music stopped. Couples began to make their way towards their seats.

She went up to him. "Dr Leggate," she said.

"I'm delighted to see you," he replied. "May I ask, please, whether we might dance during the evening?"

"I should be very pleased." She produced her programme. "Here's a waltz. I hope you don't prefer the polka. If you turn over, you will find another, there, see, and perhaps the next. It's not usual to have more than two dances with the same gentleman, but you need to catch up."

"There's nothing I should like more."

"I'm over there, with my friend Mrs Struthers and her husband, their names are Caroline and Peter. She's my closest friend, I've torn her away from her two children, Emily and baby George, for the evening, to chaperone me. I'll introduce you."

She led him among the groups of standing people, feeling his presence close to her as they walked.

"Caroline, Peter, may I introduce Dr Leggate," she said. "Dr Leggate, these are my friends Mr and Mrs Struthers."

"How do you do," said Leggate.

"Why don't you talk with Mr and Mrs Struthers," she said looking at Leggate, noticing how he presented himself to her friends. "I've agreed to dance with Mr Phipps."

The orchestra struck up.

"Then, if you can spare him," Marian said, "I've promised to introduce Dr Leggate to Mr Gibbs, Curate of St Giles, who has a formidable reputation at whist."

Reluctantly she left him with her friends, but stood for a minute to await a calming of the thudding of her heart before accosting Mr Phipps.

When Marian came to shepherd Leggate away, she said, "The sixth dance from now," and took him to be introduced to the curate.

Marian's mind was not on dances with other partners, and when the fifth ended, she went into the card room, to stand a yard or two behind Leggate, regarding him closely, noticing the curl of hair at the back of his neck, seeing he'd dressed carefully for the occasion, feeling altruistic pride.

"Really, Dr Leggate," she heard the curate saying, "I trust your skill in medicine exceeds your skill at whist."

She saw Leggate look towards Mr Gibbs, who was smiling.

"I apologize," he said. "You'd better set me a penance."

"I don't think he'd do that," replied Mrs Winthrop, who was seated on Leggate's left. "Far too Popish. More like an hour's compulsory card practice."

Everyone at the card table laughed. Marian now saw Mrs Winthrop glance towards her, which caused Leggate also to look round and see her there—his face broke into an expression of undisguised pleasure.

"If you'll excuse me," said Leggate. "I've an appointment for the next dance."

"Ah!" said Mrs Winthrop. "Now we understand. You've not been giving us your full attention."

Marian saw the little group laugh again as they released Leggate and summoned another player to take his place.

"Thank you for coming to rescue me," he said. "I'm not very good at cards."

"But you do sometimes dance?" she asked.

The music started. She stood facing him. She reached her right hand to take his left, rested her left hand on his shoulder. She felt her heart beating again, noticeably.

This state, she thought—with his physical presence, drawn to him. The closeness, as if in public we may approach just to this distance, within a private bubble just the two of us, and swirl round the room, with him, here.

And then, with a sense of profound relief, she found herself calmer, no longer that yearning—still on the edge of trembling, but a kind of right-ness.

As the dance ended, Marian said, "Thank you. Time for refreshment."

She linked her hand decorously through his arm, so that he could escort her. She sat in a chair that he placed off to one side a little, asked him to bring her something appetizing. He did so, then stood by her chair.

"Now we can talk without whirling about," she said, more comfortable now, but still unable to quell the too-strong inner excitement.

"You dance beautifully," he said. "You have a natural gift."

"You must find our provincial offerings very meagre."

"Not at all, I found the competition most refreshing."

"You're just being polite. I know you've listened to the best music in Paris, and that you write arrangements for anthems, and that you've lived in London and Edinburgh."

She noticed his slight look of surprise as she mentioned that she knew he wrote arrangements.

"I find Middlethorpe very much to my liking," he said. "Certain aspects especially so."

"You mean walks along the estuary, or to Hadby?"

"Aspects closer to hand."

"When I was in Leipzig," she said, "there were concerts on most evenings. I went to study the piano at the Conservatorium with Herr Mendelssohn. Robert Schumann and his wife, Clara Schumann, were there too, though they left for Dresden soon after I arrived. They're both wonderful musicians, though it used to pain me that he seemed so dependent on her, and he even envied her fame as a performer. If there were anyone I should wish to be like, in my piano playing I mean, it would be her."

I'm saying too much, thought Marian. Indiscreet, like my mother. I shouldn't talk about Robert Schumann fawning on his wife, it seemed so obvious to me in his every gesture when I saw them together. Why ever am I talking about it? because I fear that any moment I shall be fawning on him? shall lose myself utterly? I should be listening, not talking.

"How marvellous," Leggate said. "All these famous musicians, and to study with Mendelssohn himself."

"It was marvellous. People there are passionate about music—it's something I miss here. But tell me about yourself, when you were abroad, in Paris."

"I worked for a professor, very brilliant, Professor Magendie, it was with him that I got my interest in cholera, the problem I'm still trying to solve."

"So you had a similar experience to mine, studying with someone prominent."

"Very prominent," said Leggate. "He more or less invented the idea that we can find out how the body works in health and disease by making observations on it, piece by piece, almost like taking a watch apart."

"And you admired him?" asked Marian.

"He had his flaws," said Leggate, "he used to get into terrible disputes with people, and was always too eager for the arresting generalization, which he would cling to, quite uncritically. When we saw our first cholera patient in Sunderland he thought the man was practically dead, in a state of suspended animation. Then he had the idea: in all other diseases illness is followed by death, but in this one it was the other way round, first a kind of death, then the illness. A pronouncement. Striking you see, but unsound."

"You're not unsound, so you have the advantage of him."

She saw him look at her. She had not spoken ironically, but he seemed to be wondering if she had.

Having nibbled at the tidbits Marian rose and led Leggate towards the dance room.

"Here's Middlethorpe's musical entertainment," she said. "And I, what am I supposed to do on such occasions?"

"Supposed to do?"

"I would prefer to be among the musicians, but you see," she said nodding towards the four players, "this is the town orchestra, already complete. So what's my function?"

"To dance, converse . . . just the ordinary things."

"Hmm," she said. "May I make a small demonstration? Watch."

Marian left Leggate, went to say something to Mrs Reckitt on the other side of the room, then stood a little way apart. She extended her neck slightly, and regarded the dancers. When the music stopped, a man approached, invited her to dance.

The music began again. She danced with the young man. When the music ended, they bowed slightly to each other. She indicated a direction, towards Leggate.

"Dr Leggate, this is Mr Ashdown," she said.

"How do you do."

"Mr Ashdown works in a shipping company," said Marian. "And Dr Leggate has been in the town two years now, is that right?"

"About that," said Leggate.

Mr Ashdown made a conversational effort. "Do you like our town, doctor? What d'you think? Miss Brooks and I grew up here, so we don't know the impression it makes."

"I like it," said Leggate. "In many ways. Although I'm not out much in company, I've come to know the geography quite well."

"A fine place to ship from," said Mr Ashdown. "Close to Germany and the Baltic ports, and our estuary lets barges into the heart of manufacturing country . . ."

In a little while, after he had described further virtues of Middlethorpe, Mr Ashdown remembered that he had reserved a dance with Miss Smiley, and excused himself.

"See," said Marian. "My function's to dress up like this." She lowered her head slightly towards the left, turned her head a fraction, held one side of her skirt and made a small movement as if she were about to curtsey. "Like an advertisement in a handbill. Then I stand, poised, as if to enhance the effect of the advertisement. I needn't do anything else."

She looked at Leggate to see if he were following.

"Then some young man, who knows me slightly, takes my posture as an invitation to suggest that we dance. And if I were to become animated, he might take this as a further encouragement."

"It's harmless," said Leggate.

"But the young man, who is perfectly pleasant in himself, knows almost nothing about me. He is, instead, attracted by a cornflower blue dress."

She looked to see if she could divine what he was thinking.

"And if I were to mention Clara Schumann or Mary Wollstonecraft," she went on, "he'd have not the slightest idea what I was talking about."

"It's a way of meeting people," he said. "He might become interested in Clara Schumann."

"I don't believe so." Marian looked at him. "My function is to be decorative. I take an interest in him, and the encounter goes well, for as you see he has much to say on Baltic ports and woollen goods."

"So such conversations are barren?"

"Not at all," she said. "I find the town's trade rather interesting, but the conversations are unequal. I do have views—I think we should develop trade more with the Dutch, who will become more important now they've freed themselves thoroughly from the French and the Belgians, but no one thinks a girl's view worth listening to. My part is to be a picture in a drawing room or a statue in a garden, an ornament."

She felt herself on the edge of vehemence, though her tone had been even. I'm saying too much? she thought. But, with an inward sigh, she thought: this is me, better for him to see it now, rather than for it to be a disappointment.

"It's as if in the orchestra everyone has their function," she continued. "Mr Conductor, the violins, the French horns, the flutes, the man with the tympani. Together they create an impressive theme. If there were a woman in the ensemble—she would be beautifully dressed, of course—she might be allowed, at the right moment, in a kind of pause in the main development, to play two arpeggios on a harp."

"You'd better make a ruling then, immediately, on what we should talk about," he said.

"Hm..." she said. She saw him smiling. She felt a flood of relief. He had understood her, and he was teasing, not unkindly. It was the perfect response.

"I was going to ask about your science," she said. She found his smile so appealing that she could not help but smile back, almost apologetically. "And more about your research, but I see I've outlawed that."

"Music, then," he said. "Why is it that, though music has no meanings, it affects us so much?"

"Its patterns," she said, glancing towards him. "I think the human mind must be fashioned so that it responds to the patterning of things, and to certain kinds of patterns, especially those we call melodies."

"And it tries to discover what the patterns are," he said.

"We said we weren't going to talk about science," she said. "No—in music, I think, the notes form patterns which we apprehend quite easily, but at first in any piece of music they are either fragmentary, or they suggest variations. As we listen and hear them repeated, our minds fashion the patterns more exactly, and follow the implications that the composer has seen, and their connections, which of course with great composers are not obvious, but the patterns prompt the mind forward, yearning for completion."

"And then there comes a phrase that does bring completion?" he asked. "Does that give the meaning?"

This was a subject on which she had thought a good deal; now here was someone who not only listened, but could understand.

"In a way," she said, "but not before an interval when the listener is so taken up into a state of suspense set off by the partial patterns at the beginning, and by their elaborations, especially those that are not closed off but are augmented as the piece goes along, then as we wait for the completions the waiting is transformed into such an anticipation that it can itself become beautiful, as well as moving."

It's like people engaged to be married, she thought, with their minds full of implication, waiting in delightful anticipation, where each new action between them adds to the pattern and heightens it. This was a thought she had entertained before. This was not the time to voice it.

"So instead of a pattern coming from meanings," he said, "which happens with words, a meaning comes from the patterns."

"Something like that." Marian looked at him gratefully.

"I'm sorry," he said. "Now I'm making pronouncements."

"An apt pronouncement—the patterning can touch the structures of the mind's deepest affections, without any mediation of words or things."

"So music leads the mind inwards?"

"Robert Schumann has said music reflects the procession of states of the soul."

"And these states are our emotions?"

"Hmm," she said, aware that the conversation had swerved close to her innermost self, and that this was too soon to be speaking of such things. "That's something I wonder about a good deal."

She saw him dart a glance at her that she found wonderfully meaningful.

"But come." She beckoned him away. "Come and talk with Caroline Struthers again. I see her husband's deserted her for some affair of state, and I've said I would dance with Mr Hawley. You and I have another appointment towards the end of the evening, and we can talk some more then."

They walked towards Caroline.

"Would you care to dance, Mrs Struthers?" Leggate said.

"Just come and talk."

Marian waved gently at the two of them, a pantomimed farewell, then veered away as Leggate sat down. She drew in her breath. She found Mr Hawley, as she had promised, and during the dance kept her mind rigidly on the music and the steps she was performing, since inwardly she could scarcely contain herself. At the end, she excused herself from the company of dancers and talkers. In private, she exhaled.

Saying too much, she thought. Far too much, but it seems all right. He's paying attention, engaged in what I'm saying. I hardly know the man, and there I am ... but it feels that I can trust him with my thoughts, that he values them.

She looked at herself in a mirror, recomposed herself, went back into the dancing room, saw him there still with Caroline, and went to rejoin them.

"We've established that he's not been out much in Middlethorpe," Marian heard Caroline say to her when she returned. "Although he says he's been out doctoring and so forth, I think the sickroom doesn't count, because then only patients see him, and in any case he might catch some horrible disease."

"My friend becomes very alarmed at the idea of disease," explained Marian.

"Well, I agree," said Leggate. "It is alarming. But I have ways of keeping from catching other people's diseases."

"I'm sorry," Caroline said. "I shouldn't tax you. And you do research on cholera."

Marian saw Caroline glance at her. "We'd not yet got to that in our conversation," Caroline explained.

"It's true. I am doing research," said Leggate.

"Please solve it soon," Caroline continued. "Marian's right, disease does terrify me. I can't bear to think there even is such a thing. It's going to come again isn't it, the cholera? It can wipe out whole cities."

"It would be a very severe epidemic that killed even a tenth of the people in any place."

"I've nine brothers and sisters," she said.

"I'll do whatever I can to help you all avoid it," said Leggate.

Marian noted his kindly tone, saw his manner friendly to her friends, thought that here was someone of whom she could feel proud.

Peter Struthers returned. "We're talking about cholera," said Caroline. "Dr Leggate's subject of research. When you think . . . we're all of us at risk."

"I think you should talk about something more cheerful at a dance," said her husband. "The doctor needs an evening off too, I expect."

A little later, Marian danced with Leggate again. As they spun round the room, she felt him there, just a few inches away. So close did he feel now, that it was all she could do to stop herself putting both arms round him, and drawing him to herself, in the middle of the room, in plain view of everyone. But she did not. Instead she felt the connection between them, his hand in hers, his other hand on her back, as if these hands would make her body the lightest of objects and lift her beyond this mundane world. She felt no longer within herself, but outside—that sense again that she had experienced while playing Chopin.

The waltz ended. She stood silently beside him, for a long interval, not wanting the moment to end, not talking, watching the other dancers make their way, as if they were dream figures, back to their seats. Then came another waltz, and the same feelings continued. When it ended, she looked at him, and started towards Mr and Mrs Struthers.

"Take your partners, ladies and gentlemen," said the Master of Ceremonies. "We shall end with the Sir Roger de Coverley."

"Good night," she said. "I can't tell you how glad I am that you could come to the dance, and that we could talk. I'm going to slip away now. Peter and Caroline are taking me home."

"I've enjoyed this evening very much," he whispered.

Chapter Twenty-two

For Leggate, the image of Miss Brooks had started as a miniature, placed on the evening of the competition quietly in a corner. Within a few days the image had grown, and become a presence, pervading the whole house—the perfect wife, charming, intelligent, who would provide spiritual refreshment from his labours. That not impossible She.

I'd make her happiness my whole concern, he thought. And she? Would she not complete a life?

He crossed the room, went to lie on his couch, stared at the ceiling.

I like forthrightness, he thought. But that demonstration with that Mr what-was-his-name at the dance... Whatever did she mean by it?

He started at a blast on a steam whistle from the river.

This won't do at all, he thought. Must get on, find the evidence, before the next wave of cholera arrives.

For his experiments on flies, Leggate had obtained different kinds of food. First he put out meat in his garden shed with its door wide open. Within five minutes a small herd of flies grazed on it like contented cattle.

They must detect it from yards away, he thought. Must be attracted by its smell, I suppose. They mimic the actions of larger animals. And like cattle, I'm sure they must defecate where they eat. They eat from rubbish, then walk on human food and eat it, and defecate. Disgusting.

Leggate found what the flies would eat, observed where they would go. The different foods, meat, jam, cheese—they all attracted the animals, in different degrees.

This is a clue, he thought. Some foods are more likely to be visited by flies, hence more dangerous to eat than others.

He caught several of the insects, studied their anatomy under the low-power objective of his microscope. Their hair-covered feet, by which they could stick even to window panes, were perfect sponges to absorb the poison, and then spread it.

Having written today's observations on the flies, he turned to the article "Toxicology," in Volume IV of Forbes's *Cyclopaedia of Practical Medicine*. He read for fifteen minutes.

I'm on the right lines, he thought, poisons from plants, some are lethal, like deadly nightshade that makes belladonna. So moulds and other vegetable species that grow on midden heaps could also make poisons. Flies carry the poisons to human food, which is then eaten . . . So many links in the chain of argument. I must demonstrate each one, but I'm not a botanist, or a microscopist . . . Perhaps others have shown some of these steps.

His eyes followed the lines of the article on toxicology.

What's this? he thought. He read in Forbes: ". . . effects of poisoning by the irritants resemble very closely those which result from many forms of natural disease."

So irritants could work by perforating the lining of the gut, he thought. But the irritants are minerals. Must think about vegetable poisons.

He turned the pages. Here, he thought, "Vegetable irritants." I knew it, there are vegetable irritants too.

"The different species of ranunculi," he read, "particularly the acris and sceleratus, afford by expression or decoction with water a highly poisonous fluid . . ."

Last night he had been called out. He had not reached his bed until two. His eyes were heavy with recondite words. Lost sleep claimed him—a fitful dream: Miss Brooks, holding upon her shoulder a rose-coloured parasol, in her other hand a bunch of wild violets and dandelions. She lets go the parasol, which floats into the air. She takes from her bunch a dandelion globe of tiny parasols. She blows gently, once, twice . . . "He loves me, he loves me not . . . ," with each phrase a cluster of the little seeds floats into the air. As each one comes to earth, up thrusts a vigorous green leaf, that transforms into a dandelion, around it a golden haze of pollen specks, each of which contracts into another perfectly formed pink globe of tiny parasols.

He woke, disoriented. All these plants reproducing themselves, he thought.

He went down to the parlour, and picked up the copy of *The Middlethorpe Examiner* that lay on the table. "Trail of clues leads to murderer," said the headline. "By our special correspondent."

Last week, by tracing a minute trail of clues, the identity of the murderer of Marie Watson became known,
due to the tenacity of a gentleman who prefers to remain
anonymous, working in cooperation with the police . . .

Attending Connie and her baby at the house in Friars Walk, Leggate had
heard that the special correspondent was the gentleman who preferred to
remain anonymous—Edward Warrinder—and that he had taken up with
Millie, one of the women in the house.

Not all the women liked Warrinder. "There's a good few of his sort,"
Amy had told Leggate. "They want us for companions, and for you know
what, then we're nothing."

"Look at the way he writes," Liza had said. "When he's here, you'd think
he was friendly, but when he writes he's all sneers."

Millie had defended him. "He treats me quite fair, and he hasn't hit me,
and his drinking has its good side. He falls asleep and I'm off duty."

The other women had laughed.

Leggate continued reading the newspaper article.

The trail began in a house in Friars Walk, a notorious
place, the residence of the murdered woman who had
come to this country from the West Indies when she was
seventeen. Brought up a Roman Catholic . . .

The article described how the police, and the gentleman who preferred
to remain anonymous, had almost given up in frustration after weeks of trying to spot the suspect in likely haunts in Leeds, where he was believed to
live, until:

With the help of Watson's associate, the gentleman went
through every last article of Watson's belongings in the
room she had occupied. Perhaps a keepsake would be
sufficiently distinctive to take to a jeweller, perhaps some
item bore an inscription. Thus every piece of clothing,

every geegaw, every article of toilette, every comb, every camel-hair brush, every receptacle of rouge, every delicate vial, every item used by such women in the course of adornment was inspected.

The survey of the deceased's belongings seemed without useful issue, until the gentleman investigator arrayed all the smaller items on a table, for what purpose he was not at first able to tell with any exactitude, but this was the chance that finally gave the indication that was needed, not immediately, but when he came again to view this array the day after he had set it out.

The realization came; the vital clue stared him in the face. It would have escaped anyone whose mind was not attuned minutely to these investigations. There were ten vials of scent, of different sizes, of different perfumes, but from the same manufacturer. The notion struck him that most women wear but a single perfume. This woman had ten different kinds.

"Yes, women usually wear one kind," the gentleman's informant told him. "At least if they buy it themselves. But Marie did not. I sh'ld think these was given to her."

Were the different perfumes given by different men? Perhaps; but if there were but one purchaser, how unlikely it would be that he would buy different perfumes but all from the same manufacturer. Shops sell the wares of many manufacturers and importers.

"Perhaps then," the thought struck the investigator, "the man was himself the manufacturer of these perfumes or the importer. Perhaps his name was on these very bottles."

Leggate had heard how Millie had accompanied Warrinder once more to Leeds secretly to catch a sight of the perfumer entering or leaving his place of business, to see if she could identify him. They had stayed, as Millie recounted, for rather longer than might be judged necessary, at the best hotel in Leeds, all paid for by the newspaper.

"It was him at last," Millie had said, with some pride. "That used to come here. Turns out his name's Augustus Biggs, and in the newspaper next week you can read about the arrest."

Warrinder's already tried this man and found him guilty, thought Leggate. The town hangs on his every printed word. A cockerel strutting in a farmyard, with a topic to suit his talents. A relief, though, that he's got the man; when he eyed me in that insolent way in the house in Friars Walk, I thought for a moment he was about to start concocting something about me.

Why waste time with this? he thought.

He threw the paper down, and went to tend his experiments on flies in his garden shed. He was there for twenty minutes.

He gently scraped the surface of the foods on which flies had walked, mixed some scrapings with water, some with alcohol, to make extracts. Finally he brought his samples indoors, back to his study.

So, he thought, this is a step forward; if I scrape the poison not from dung heaps, but from food that's been left out, that will be a step closer to the people who would eat it. When the cholera comes, I'll make an extract of the poison from the food, then apply that to the gut of a rabbit freshly dead, and inject mercury into an artery of its gut, to see if it becomes permeable.

Leggate tried looking down his microscope at the scrapings from the food on which flies had walked, to see if particles were visible, but everything was obscured by the patches of mould that had started growing on the food— variously brown, grey, reddish. All that seemed certain was the smell he had brought into his study.

Leggate rose from his table, and looked out at the river. For Warrinder it's all clear, he thought. But there's nothing to publish here. Whatever the flies carry to the food doesn't make the gut permeable except when the cholera prevails, so there's no point in trying that experiment now. But I'll get in mercury ready, and get some syringes. I'll go for a walk, to clear my head.

Beside the river Leggate walked northwards; outside the town beyond the new waterworks, he sat on the grass. Insects buzzed between the plants, the wretched flies and other creatures, all about their business.

He picked a dandelion globe of tiny parasols, idly started puffing: She loves me ... She loves me not ...

He could not bring himself to finish the game of alternating puffs and phrases. He did not want to tempt the possibility of discovering that she did not. He lobbed the dandelion into the stream. How does the French one go? he asked himself. *Je t'aime un peu, beaucoup, passionément, à la folie. C'est ça?*

He picked another dandelion. Its globe was distended: *Un peu ... beaucoup ... passionément ...*

He blew gently for the fourth phrase, "*à la folie,*" hoping some parasols would be left to repeat the cycle, past the safety of "*un peu,*" perhaps as far as "*beaucoup.*" The game required that in each puff at least one parasol should be freed. He misjudged, and all the remaining parasols were dislodged.

Folie—madness, he thought. Just a child's game ... Margaret made enquiries ... Miss Brooks is a respectable young lady. It wouldn't be madness, not at all.

Plucking another dandelion, Leggate blew a prolonged gust to empty his lungs, and watched the swirl of parasols drift in the summer air, to mingle with insects that buzzed above a cow pat.

Each species has a means of spreading itself, both pollen and seeds, he thought, some in the air, and some by insects. First the moulds grow on dung heaps. The patches of mould that we see are clumps of thousands of tiny plants, and like the larger plants each one has its seeds that are too small to see, each plant must bear thousands of seeds, so the flies walk among these plants as bees walk through pollen, and the invisible seeds cling to their feet like minuscule burrs, then they take them to human food ...

So it's not a poison, or not directly, it could be invisible seeds!

. Leggate's mind was in motion: seeds are eaten, grow in the intestine, he thought. Their roots penetrate, seeking moisture from the blood, making holes in blood vessels ... could that be it? That's how vegetable irritants could work, their roots bore holes into the membranes. What could be more irritating?

Leggate rose from the river-bank, excited, in concentrated thought. It's not a mineral substance, it's seeds. The seeds of moulds that stick on the feet of flies are deposited on the food—that's why all those moulds start germinating on the food—and when swallowed the seeds germinate in the intestine too. Once planted they grow in profusion . . . cholera is spread not by touch or miasma—which are immaterial—but by tiny seeds. That's how such a small amount of poison has such immense effects, seeds that reproduce themselves. When the cholera strikes again, I can examine intestines post-mortem with a microscope, see tiny plants growing, they'll be like the moulds. No one has thought to look in the right way with the microscope. Not a mineral poison at all, but seeds which multiply.

"That's right," he announced aloud to the river. "And therefore . . ."

He glanced round to see that no one had heard him, and saw he had walked to the waterworks with its tall hydraulic tower, the striking new landmark in these parts.

A single mistake all along, he thought. Everyone's made it, to think of transmission as a poisonous substance, a mineral, but it's not a mineral, it's seeds!

He walked farther north, thinking, thinking, fitting his new idea of seeds into everything he knew. Everything fitted. This is what had been missing—this is how so little poison can have such large effect. Now it makes sense, just as a small acorn grows to a large oak, and a sackful of grain plants a whole cornfield. No one's seen it before.

Only when the buildings of Battersby came in sight did he realize how long he had walked. He turned, and retraced his steps. When he came again to the waterworks on the edge of Middlethorpe, his new conception was in place. He no longer felt agitated; instead he felt an inner calm, a satisfaction. The problem, so elusive to so many, though not yet solved, could be understood.

I'll go home and write all this down, he thought. He took out his watch. Dinner will be soon. I'll do it after dinner then. I've got half an hour now, why not go and look at the water tower, and the steam engine and pumps, that everyone's so proud of?

The town council was indeed proud of the scheme, and still had its workings open to the public. Indeed they employed a man who would take visitors to see its several aspects.

Leggate entered the building. There was the wonderful steam engine, its reciprocating piston, its glistening rods and beam, its smoothly turning wheel, and the two great pumps that forced the water upwards to the reservoir at the top of the tower.

He climbed the tower to see the town spread out before him, and beyond it the estuary. To the north the river wound its way, where he had just walked, towards Battersby. Before going down, he went again to the south side: yes, there, running across a flat field from the foot of the tower, was the great twenty-two-inch pipe that took water into the town.

Just like the body, thought Leggate, as he walked down the tower steps. Pump and pipes. Exactly alike. And cholera would be the pipes that go up the tower and into the town, becoming suddenly porous with some giant mould with roots that would bore into the metal, or perhaps would turn the metal quickly into patches of rust, so that it was perforated with a thousand tiny holes. One such affliction, and none of it would work.

After dinner Leggate went to his study, to his notes of his new idea. Late in the night he finished what he wanted to write.

I've not got the evidence yet, he thought, but I know this is right. The spread of disease is like the propagation of plants, and the little planticles grow in the intestine, and bore their roots into the blood vessels. I feel it in my bones. The microscope—that'll be the thing. I must talk to Newcombe.

That's the meaning of my experiments with flies, that's why all those moulds grow, and that's how there's sufficient of the substance to make people ill, because of microscopic seeds that propagate and multiply. It only needs a few seeds to be eaten, maybe just one which germinates and reproduces itself in the intestine, which is fertile soil for them, and it would explain why both the fertility of local conditions and something from abroad are important.

If challenged, I shall say, "For the present we surmise that there are tiny plants, some of which are poisonous. Those present now are not the seeds of cholera, but of some other species, which gives people diarrhoea, though much milder. The cholera planticles aren't growing here at present, and that's why there's no cholera. But I make you this challenge. When the cholera returns, I will keep my foodstuffs covered—it is a simple expedient. What will you do? Will you be prepared to leave your food exposed to the flies that carry the seeds? Then we shall see which of us contracts the disease."

Covering food, he thought. So simple. There's no inconvenience or harm. I shall advise people to cover their food, and then, when the epidemic has passed, I shall count up those who took my advice, see who did and who did not contract the disease. *Voilà*.

Leggate's thought of sending off something to *The Medical Gazette* was to have it accepted immediately, published in a week or two, and be received enthusiastically.

Halliwell would say the seeds are still just ideas, he thought, moderating his excitement. I must get the evidence.

He wrote his planticle theory in his notebook. After his notes he began a copy in his very best hand, heading it:

FOR PRIVATE CIRCULATION ONLY

I'll make three copies, one for myself, one for Halliwell, one for Newcombe, he thought. In my will, if the disease returns and I should succumb, I shall leave money to Newcombe to appoint an editor to go over it, and have it published.

Thus Leggate stilled his mind over the question of priority, of how he might deal with having found the right answer, but not yet the evidence to convince a sceptic.

Chapter Twenty-three

Marian mused on the events of the Larkins' dance. I shouldn't have made that demonstration with Mr Ashdown, she thought. Now I've put off one of the few presentable men in Middlethorpe.

Her mind reached back to Leipzig, to Andrei, the Russian cellist. Here I am at the same point again, she thought, believing a man is interested in me... I've not got that feminine softness that Caroline has, that the wretched Grete had. And I'm growing old, soon it'll be too late.

Marian's aunt had impressed much upon her. So indeed had Mary Wollstonecraft, whose principles were poignantly illustrated by the spectacle of her own mother who lacked all companionship in her marriage.

Here I am, she thought once more, again wondering how to engage the male affections.

She took up Mary Wollstonecraft, the book that, since she had first opened it, seemed so perfectly to put into words so many of her own thoughts on this topic. She found a passage she had marked.

"Love, the common passion," she read, "in which chance and sensation take the place of choice and reason..."

Is that what's captured me? she thought, chance and sensation? which in a moment obliterate all I've believed?

She read on: "... in some degree, felt by the mass of mankind."

So at least I'm not unusual, she thought. I'd started to believe my inability to experience such sensations towards anyone who was the least bit possible showed I was perverse.

Marian read the next paragraph: "Friendship or indifference inevitably succeeds love."

Friendship, she thought. Can we imagine the empty-headed Grete developing a well-rounded friendship with her husband? But it's everyone, my mother, Caroline, if they all settle for indifference, all of us women...

"... made ridiculous and useless when the short-lived bloom of beauty is over," she read.

"That can't always be how it goes," she said aloud to herself.

Mary Wollstonecraft married, she thought. We don't have to be ridiculous and useless. The other must be possible. He understands me, with him we could be friends.

Marian said to her aunt, "I feel something strong towards him. Though how does one know about such a momentous thing? And I think from some signs that he likes me. You were in love ... but you decided against ..."

"I said that in the marriage service I would promise to obey, on the condition that he would say privately that this would not bind us. We talked about it a great deal. He would say that ordinarily it would not come to that, that this would only be a final resort. I said I was sure he could find someone more pliant."

"And you've not regretted it?"

"He did find someone more pliant. And of course I've regretted it."

Marian discussed with her mother the plan of inviting Leggate to dine, and wrote on the back of the card:

My mother would like me to play the Piano for her guests after dinner. I thought it might be pleasing to have, in addition to something instrumental, one or two Vocal Pieces that I could accompany. I know you sing, so if you would call, say this afternoon at about four, we could discuss the possibility. If you agreed, you could borrow the music for a day, then we could practise.

At four o'clock Leggate arrived at the Brookses' house, was shown to the drawing room in which sat Marian with her mother, at their embroidery. The plan for two songs was discussed. All three went to Marian's music room where she played the accompaniment of the songs, as Leggate, looking over her shoulder, read the words.

When he left, he took the music and an appointment to return next day.

Although it is not considered proper for a single lady who is not engaged to spend time alone with a gentleman, the convention evidently does not apply to musical practice. So on the next day Leggate and Marian passed an afternoon unchaperoned. During part of it he sang, accompanied by her. During part she played the piano for him. During part they talked about the songs they had practised, during part they spoke of the time she had spent in Lausanne, and in Leipzig.

She listened as he told her about his researches: "I am making some experiments."

"More experiments?"

"Have I spoken of experiments before?"

"When you attended my father you mentioned experiments in Paris on effects of letting blood."

"These are new experiments, on how diseases spread."

"By contagion, you mean?"

"People have thought that cholera spreads by contagion, as you say, or by something in the atmosphere that we breathe. But it's not so. It's something we swallow, not really a poison but invisible seeds of minuscule plants."

"So you need a microscope to see them?"

"Exactly."

"And you've made this discovery?" asked Marian.

"Almost, but I need more evidence," said Leggate. "It is not so easy. Professor Magendie, for whom I worked, does experiments on everything."

She looked at him intently.

"One has to make a discovery," he explained, "that no one has made before, then be first to the press."

"Like a race."

"No one knows the route, or which idea will be right, or what exactly to try. You only publish when the whole train of evidence is complete—if you make a mistake, there's nothing but ridicule. You'd be considered unsound, and never taken seriously again."

There was only one discordant note: "One link in my chain is flies," said Leggate. "I need to show how they spread the invisible seeds from dung heaps, the seeds that cause the disease."

"Dung heaps . . . ?"

"The experiments will soon be finished," he said abruptly.

Marian was surprised that the rush of talk stopped so suddenly. Surely he's not embarrassed at hearing himself mention dung heaps, she thought.

"Don't stop," she said. "I like hearing you talk about it, please say some more."

"One of my next steps is to try to describe exactly what medical advice to give people."

Marian heard from his tone that the topic was finished. She wanted to continue. "You were explaining how the disease spreads, by the seeds."

"It's not entirely clear yet. I'm writing it up for publication, and when it's in that state it'll be much clearer," he said, "and I shall be able to show you."

She saw in his eyes that something, some apprehension, had stopped him, that now he was trying to reassure her that no breach had occurred.

"What about Schubert?" he asked.

What's wrong? she thought. Why won't he discuss it? Is it because it's not yet fully thought out? He could use me as a sounding board. I could be of help to him.

But she fell in with the new topic.

As Leggate prepared to leave, she said: "I hope you approved the pieces I played for the competition." The moment she spoke, she regretted soliciting a compliment in so brazen a manner.

"I approved both choice and performance. The Étude and the Prélude are very great pieces of music, which require the utmost musicianship, and the Nocturne, which you played at the end, though not as grand, affected me very much too."

"I am glad you approved."

Marian and her mother had together decided the arrangement of seating for the dinner party. Colonel Willoughby, who had been in India, could sit next to Mr Brooks. Although this placed two gentlemen beside each other, it would keep both content, for they could discuss campaigns. Marian would sit on the left of Dr Leggate, Miss Botham on his right.

As dinner progressed, and after he had conversed for a decent interval with Miss Botham, the time came for Leggate to talk with Marian. He began to speak of how scientists were much respected in Paris, in London, in Edinburgh.

"Sir Charles Bell was a fellow of the Royal Society in London and the one in Edinburgh, and of the Royal College of Surgeons," he said. "He did research on nerves parallel to work done by my professor, though Magendie always maintained he had the priority. They had a terrible dispute."

Leggate sucked in his breath. "Bell received a knighthood in recognition of his researches in anatomy and medical matters," he said.

"Indeed," replied Marian.

There was a pause, during which she thought: is this what scientists aspire to? Knighthoods? Are they inwardly so fragile that they need such accolades, while all the time they talk of truth!

Fearing she had been abrupt she said, "How did you choose doctoring? It's not such a usual pursuit for a gentleman."

"When I was nine, I think. My brother died, and there was a doctor who attended."

She saw him regarding her. After talk of the Royal Society and of knight-hoods, he seemed suddenly vulnerable, as if this were a topic that touched him closely.

"You felt close to your brother?" she said.

"He was two years older..." said Leggate, looking pensive. "The doctor came often, and did everything he could, he must have noticed how fright-ened I was, and then when my brother died..."

Marian regarded him sympathetically.

"The doctor would call even after that, and would say: 'To see how you're getting on, young fellow'—extraordinary now, to think of his kindness."

"And you saw that here was something worthwhile to do in one's life, and help prevent what had happened to your brother?"

She saw him looking at her, his eyes searching hers. A chord of sympa-thy had been sounded. She was moved. At the same time she knew this was something that touched him. She feared that if she should ask more, he might be too affected by recollection. The dinner table was not the place.

"Then to Paris, to take up research," she said, guiding the subject towards something safer. "That was where you worked with this Magendie, who had the dispute with Bell."

"With Magendie," he said, "that's right. To exchange the dubious crafts of medicine for something solid, for science."

"You sound like a believer," she said, secretly exhilarated that here was someone with such a passion as might fill a moral void. "To work not just for the few, but for everyone."

She saw his eyes again, looking at her as if grateful that she had put into words what was in his mind.

"It was very important to me," he said. "I was one of the first to see cholera in England. He took me with him to Sunderland, as his assistant, at the end of 1831—I was still very young, I was with him every day for nearly three weeks. He was wonderfully scathing about medicine as most doctors practise it. I thought his idea that disease could be understood like the workings of a machine, and could therefore be conquered, was the most important thing I'd ever heard. Without quite knowing it I'd suddenly become part of a new movement. More exciting than could be imagined."

Marian noticed that after dinner when the men came to rejoin the ladies, her mother made a point of talking to Leggate. Marian joined a little group that consisted of Colonel and Mrs Willoughby, Mr Jinks, and Miss Botham. She stood where she was able to overhear her mother's conversation.

"Before you came here, you were in Edinburgh?" said Mrs Brooks to Leggate.

"For several years, completing my medical degree, and working as assistant to a physician in the city, Dr Guthrie," he said.

"Ah," said Mrs Brooks. "But you're not from Edinburgh?"

"I was brought up in a village, well, no longer a village, part of London almost. It is called Dulwich, and I went to school there, to Alleynes."

"And your family's from Dorset."

"Near Bridport."

"I see . . ."

"My mother is French. I speak French almost as well as English. I began

my medical studies in Paris, working with a French Academician, Professor
Magendie."

"Your mother and father are both alive and well, I hope."

"My mother, yes. She still lives in Dulwich, but my father died, a few years
ago."

"You have my condolence."

Soon the musical performances began. Marian played one of Mendelssohn's
"Songs Without Words." Then Miss Botham sang to Marian's accompaniment. Last, Leggate, accompanied by Marian, performed the songs they had
practised.

The little group applauded warmly. Applause after musical performance
was not new to Marian, but now as she rose from the piano stool, as she saw
her mother smiling—the performance had gone well with piano and voice
complementing each other—now as she stood next to Leggate, curtseying
as he bowed, she suddenly experienced the association between them declared
and commended. With a glance again at her mother, who again smiled at
her in what seemed an approving way, Marian felt the beginnings of a blush
spread upwards from her neck to skin behind her ears.

Leggate dropped in at Mrs Brooks's next Thursday-evening gathering. He
had found an edition of Schubert *lieder*. He brought a calfskin music case,
in which he put the Schubert.

Leggate was shown up to the drawing room. Mrs Brooks greeted him.
"Come in, doctor. I'm pleased to see you." She took him, intimately, by
the arm.

Leggate glanced round the room.

"Mr Brooks isn't here this evening," said Mrs Brooks. "He's off in York.
An ecclesiastical function. I apologize."

"I am pleased to be here," said Leggate. He glimpsed Marian in conversation with Warrinder. He suppressed a pang of annoyance.

"Come and talk to my sister, Miss Mitchell. It'll be good for you two to
meet somewhere other than the sickroom."

"How do you do," said Leggate. "I'm glad to see you looking so well."

"Is that a medical greeting, doctor?" asked Miss Mitchell. "Come and sit by me."

"I'll leave you two together," said Mrs Brooks. "And don't dose him up with radicalism, remember he has to make his way in the town."

Marian's aunt looked at him quizzically. "Tell me, what do you think of my niece?"

Confronted so blatantly, Leggate was unable to reply.

"Is my question inappropriate?"

"Not inappropriate," he said. "I admire Miss Brooks . . . very much, her skill at the piano, she's enchanting."

"You make her sound as if she dispenses magic spells."

"I didn't mean that." He looked at Miss Mitchell, who was regarding him not unkindly. "I do admire her," he said.

"You are right to do so," said her aunt. "Here she is now."

Marian had detached herself from Warrinder. She now approached Leggate and Miss Mitchell. Leggate rose, and took her outstretched hand.

"How do you do, Miss Brooks."

"We were just speaking of you," said Miss Mitchell. "Tell me," she continued, addressing her niece, "are these the usual people your mother sees on Thursdays?" She looked towards Leggate to make him an explanation. "I seldom come to these events."

"My aunt thinks my father suspects her of being an atheist," said Marian. "She comes to the house less than I would like."

"Is that true, Miss Mitchell, that you are an atheist?" asked Leggate.

"It doesn't do for women to think for themselves," she said. "I'll explain the proper form. First you tell me what you think, then Marian and I will agree with you." Miss Mitchell and Marian exchanged looks.

"I think that if God wrote the book of Nature, he made it very hard to read," said Leggate.

"And we agree," said the aunt, making Leggate laugh. "See how easy it is."

"I'd like to talk," said Leggate. "But it must be some other time. I have something urgent, I am afraid, and I can't stay. I just dropped something in for you." He looked at Marian. "We had to practise some songs together," he explained to the aunt. "I was swept in by the servant, I only meant to leave some music for Miss Brooks. It is on a table, in the hall."

"Then let's retrieve it," said Marian. "Will you excuse me for a moment, Aunt. I shall be back directly."

Leggate and Marian rose. "Goodbye, Miss Mitchell."

"Goodbye, Dr Leggate. I hope when next we meet you give me a little more of your time."

"I promise I shall."

In the hall Leggate found the music case. "I thought you might like it, if you have to take music out anywhere."

"It's most handsome." Marian seemed pleased.

"I shall be able to think of you, think of it with you," he said.

Marian opened the case, retrieved the Schubert. She smiled. "Why thank you. It is a lovely present, the music, and especially the case."

Marian returned to her aunt, having seen Leggate out without summoning the maid, having stood reflecting for a minute, having then taken the music case to her music room.

"What do you think?" Marian said.

"I think you have been away far too long, letting your beau out, and I think I'm very indulgent towards you, since the rapidity of his departure bordered on the impolite."

"Was I very long?"

"Very long. I also think that he's in love with you, and it doesn't need much acuity to see that you're in love with him."

Marian heard her aunt's words with a relief that surprised her.

"But you don't disapprove?" she said.

"I disapprove less than I would many of the young men I know."

"Aunt, please . . ."

"I'm content with him as a doctor. He's thoughtful. And he shows respect. Thoughtfulness is good, though respect is rarer. But you—you must beware that with your theories you don't exert yourself too much to mould the world to your pattern."

Marian did not hear the warning her aunt offered—she was more intent on moulding her. "But you like him? I can see really you do."

"That is for you rather than for me. My question is: should I approve him and stay your friend, or disapprove and we should fall out?"

"Oh, Aunt." Marian, who had remained standing since her return, now bent down and kissed her aunt tenderly on the cheek.

Marian recounted to Caroline and how she had felt touched by his present of the music case.

"You think about him in that way?" Caroline said. "You must be in love."

"It seems much as one reads about," Marian said. "I think of him too much, I make plans for us together, I find myself looking forward to seeing him, and when I'm with him, gazing at him. Being alone has come to seem arid."

"These are the symptoms." Caroline put a hand on the forearm of her friend. "I'd despaired of you. But how good that he's a doctor. Just think of all those consultations so readily available."

"Those would be what you would like. I'm surprised you didn't marry a doctor."

"There wasn't one for me to consider," said Caroline. "But perhaps you'll let me consult him from time to time."

"It lifts a weight," said Marian. "Not a weight. He fills an emptiness, that only music used to fill. Is that how you felt?"

Caroline did not reply.

Chapter Twenty-four

Marian wrote a note to Leggate a few days after she had received his present of the music case. A concert party was being got up several weeks hence. Would he be able once again to sing to her accompaniment?

"Of course, we will need to practise," she wrote. "How about some of the Schubert songs you gave me? Could you call this evening, so that we could arrange it."

Their practices began.

"I could sing this, 'Where is Sylvia?'" he said. "It's in English." He read the last part of the middle verse.

Love doth to her eyes repair,
To help him of his blindness,
And, being help'd, inhabits there.

She saw him scrutinizing her. She covered her mouth with her hand, not knowing how to reply.

"Very well," she said. "We can do that. It's got an exciting piano part, and the melody for the voice blends beautifully. And this one, 'Ständchen.' You should sing that in German?"

"My German is execrable."

She broke into laughter, which dissolved the anxiety she felt.

"You doctors," she said. "You use such funny words. I'll translate the song so that you can get the sense. All these songs are about the emotions. Schubert's genius is to suggest an emotion in a phrase, and then in the next phrase transform it. I'll help you. But the German does sound better. *Laß auch Dir die Brust bewegen.* It means 'Let your heart, too, be moved.'"

Each morning became a prelude to an afternoon spent in such ways, each evening a coda of intense remembrance.

Leggate worried that he did not have enough time for his article. But for him the days took on a new quality. Instead of research providing the structure of his existence, supporting reading, note taking, and patients, these more habitual matters became an almost dreamy pastime filling the gaps until another afternoon would come, when he would be with her.

He wrote his notes, never worrying whether a comma was misplaced, but he did not get much further with his paper. For in an article, a word well chosen or ill makes the difference when the work is exposed to public view. Such an exacting task, depending as it did on summoning to his mind the consciousness of potential flaws, of potential criticism of every word he wrote, did not fit his mood.

It's natural, he thought, that I should be concerned with her just now. It's extraordinary to find someone so educated—and not just her piano playing, but someone who understands the importance of my work. With her by my side, should I not work better? I would do it for her sake.

In a week or two, my experiments on flies will be finished, now I know the flies carry the seeds of all those moulds, and I'll just do a first short article, as Halliwell suggests, giving details of my alimentary theory, then a second on the planticle theory. I'll do a monograph showing the course of the whole investigation, including epidemiological maps with the distribution of deaths in 1832, locating the middens and other nuisances, and make the suggestion that flies can probably just travel over a short distance each day. In a longer work, I need not be so careful with *le mot juste*. That's the way forward. Being with her would help me to think. I could do it the more easily for her. I shall dedicate the monograph: "To Marian."

For Leggate the principal question was no longer about disease. It had become: If I were to ask her, would she accept?

Marian sorted out her music, one afternoon, as they prepared to practise.

"Tell me about your brother, who died," she said.

"He was two years older. I suppose I admired him," said Leggate. "He always included me in things, looking back, and having seen other boys, I see now how generous that was."

"And you missed him terribly."

"Hm," said Leggate. "It's a bit of a fog, my mother was very affected, I don't think she ever really recovered, perhaps not my father either... but there I'm less sure, my father was always more distant."

"Distant?"

"Nothing I could do was ever good enough for him. He often made me feel very small."

Marian looked at him contemplatively, sympathetically, but did not say exactly what was in her mind; instead she said, "So as well as losing the brother you loved, you lost your mother?"

She saw Leggate stare at her, wide-eyed.

"I've been spared losses of that kind," said Marian. "My mother ... she's very forthright, but affectionate ... I couldn't have a better mother."

"I like her very much," said Leggate. "But how did you become so educated. Girls don't read, or know so much, and you play the piano too."

"My father," said Marian. "He decided his daughter shouldn't merely tinkle on the piano, but should master it. He was an organist, a gifted one. He taught me for four years, then passed me over to Herr Hoffman—but still made sure I practised."

"But you're educated, more so than men who've been to the university."

"He thought a tutor could as easily instruct two as one. William's two years younger than me, he was never very apt in the classics, but pleased enough to have a sister who would do his Latin scansion, and who would keep the tutor's attention largely upon her."

"So you had tutors throughout your childhood?"

"Until thirteen or so, and my father made us write. 'There's nothing as important as being able to write properly,' he'd say to the tutor. 'Have them each compose an essay of three pages on the fortifications of Middlethorpe, show them the walls, and explain it. Look at their first efforts, and correct them. When they have learned from your remarks, they will each write a fair copy for me,' so that's what we did."

"I wish I'd been taught to write," said Leggate.

"You said you had an inestimable experience in science," she said, raising her eyebrows, in mock remonstrance. "Tutors came to an end pretty soon though, my father started to find me very tiresome, and I was sent to school at Miss Davignor's—and the curriculum changed from Latin to polite conversation in French, and the art of entering a carriage."

"But you didn't give it up—education?"

"I have a secret resource: an aunt with rooms full of books and literary friends in London."

Marian would pick up the newspaper each week. She still felt guilty at having promised always to be Warrinder's friend; perhaps this contributed to her interest in his writings. Warrinder seemed to be at no loss to keep his readers' interest in the case of Marie Watson in the weeks before the case came to trial.

THE SCIENTIFIC DISCOVERY OF MURDER

Englishmen might well blush that wanton crimes continue in our land, crimes of a kind that, in their melancholy horror, almost defy description. Among them murder, such as the recent case of Marie Watson in our town, for which Augustus Biggs is soon to be tried, must surely be the most repugnant, the more so when committed in a demonstrative way, not by a person already sunk in depravity, but by a man who holds some position. But if we hang our heads in shame with the knowledge that such deeds occur not in benighted lands beyond the seas, but here in the very bosom of a civilization as advanced as ours, we may allow ourselves one consolation. Among our contemporaries are those who expend all efforts to pursue the wicked. If a guilty wretch believes he shall escape by flying on the wings of steam at thirty miles an hour, he is overtaken by a swifter message, travelling at the speed of lightning by way of the electric telegraph, which can send an account of the crime, and a description of its perpetrator, to the remotest parts of the kingdom. The case of Tawell, for instance, who poisoned his female associate just two

years ago, showed just how decisive the benefits of science can be . . .

He makes a little go very far, thought Marian. One would think if there
were nothing to say he would keep quiet, and write instead about new devel-
opments in the port, or the building of a new ship, but perhaps that's not
what people wish to read.

Marian and Leggate, in those weeks of practices for the concert party, gave
each other a hundred tiny evidences. Towards the end of the practices, one
day they were in Marian's music room, having conversed, having performed
Schubert's *"Ständchen"* until its emotional nuances were almost perfected.

When their third rendition of the song ended, Marian turned to look at
him as he stood beside her in position to read the music and turn the pages.

"Isn't music the expression of the soul?" she said. "I feel I make myself a
conduit, so that the music flows through me. It doesn't happen often, but
when it does it's unmistakable. As if the composer has created something
which exists on some other plane, and I have to allow it to reach this one."

He knew she was saying something not previously said to anyone. He saw
her aspiration, saw her striving for a perfection beyond the mundane.

"It's as if one longs for something perfect," she said. "And then by means
of brain and fingers, this thing that one desires actually comes to exist, and
the mind reaches out to apprehend it—the yearning of imagination is
matched exactly by what is."

Her words stirred an echo in his mind. "A kind of revelation," he said.

She saw him looking at her in a way that was full of meaning, felt a strong
sensation not in her but between her and him, felt a presentiment of something important, to which her heart assented.

She returned to the keyboard, with him still standing slightly behind her
left shoulder, and said, "Listen to this. A Prélude, in E Minor."

It was Chopin, a prélude he had not heard before, as if in illustration.

She had prepared it especially for him. The inner voice of its opening left-
hand minor chords, accompanied by the right hand playing single notes, gathered into a span all the sadness of his previous loss, moving him so that tears

started in his eyes. Then, within a phrase, the melody endowed him with a hope more solid than the room in which he stood, and he felt the connection between the two of them. Desire and actuality matched.

Perhaps the music contributed, perhaps the closeness of her neck, perhaps the memories recently revived of Evette—though many would think these inappropriate for importation into an English music room. The piece was no longer than two minutes. With its final chord, his face descended to kiss her neck. Her right hand left the keyboard to rest upon his head.

She rose. They faced each other, each holding both hands of the other. They agreed that he should, that evening, approach Mr Brooks to request his daughter's hand.

III.
Andante

Discovery, Marriage, Disclosure

None but those who have experienced them can conceive the entice-ments of science. In other studies you go as far as others have gone before you, and there is nothing more to know: but in a scientific pursuit there is continual food for discovery and wonder. A mind of moderate proficiency which closely pursued one study must infallibly arrive at great proficiency . . .

Mary Shelley, Frankenstein; or The Modern Prometheus

Chapter Twenty-five

It was not a remote inference for Marian to know her father would be displeased when her mother broached the idea of her wishing to marry. His insistence that she contemplate the courtship of Willoughby's repellent son, which remained in abeyance while she seemed set on spinsterhood, would certainly be brought to bear.

"He's worried that Dr Leggate has advanced views, like my sister," her mother told Marian, as if this would explain Mr Brooks's attitude.

"You can bring him round," said Marian. "You like Dr Leggate, you said you do."

"I do like him, and I'm very pleased," said her mother. "More than pleased, I shall be proud to have him as a son-in-law."

Marian knew her mother was unused to prevarication. Her usual tendency was towards too much directness, and to an alliance with herself. Now this tendency was reasserted.

"So, tell me then, what do you suggest I say to your father?"

"Tell him I'm a terrible bluestocking. Say I'm past the age of offers from acceptable men, that this will be my last chance before becoming an old maid and being stuck in his house for ever."

"I shall say nothing of the sort."

"Well..."

"He'll attempt a rearguard action," said Marian's mother. "He'll say that Willoughby's son is in a very good regiment."

"I've not the slightest interest in regiments."

Marian saw her mother look surprised by the scorn in her voice.

"I don't mean Grandfather," said Marian. "I'm sorry, but really, Mother! Willoughby's son, you know how I feel about him. You despise him yourself. He's a toad. Would you want him for a son-in-law?"

"I dare say you're right."

"So I can leave it to you?"

When Dr Leggate arrived that evening, therefore, rather than meeting with Mr Brooks alone, he found himself with both parents.

When Marian was called in to drink a toast, her father was barely polite, but she was pleased that after the toast to their forthcoming marriage, her mother gave Leggate the warmest of kisses on his cheek.

"There," she said. "It's decided, and toasted. And here's to you." She raised her glass once more. "To my new son."

With their engagement, Marian and Leggate could meet more freely. She played for him another new piece. "It's nowhere near perfect," she said. "Let's not talk about it now. What of you, and your researches?"

"It sounded perfect to me—I wish I were as close to my goal. I feel I am approaching the truth of how diseases are communicated, but there's still some way to go."

"Well, tell me."

"When my experiments are done, I was thinking I might ask Newcombe to help me, with the microscope. He's an accomplished microscopist. It's possible we might be able to see the tiny seeds which I think may carry the cholera—I told you about them."

He talked about his idea . . . Marian listened, excitedly. He's immersed in it, she thought, utterly. When he talks, it takes him over, he's inside it.

A fine mind in motion, she thought. What could be better?

She had another thought: shall I be able to distract him, draw his thoughts away to the ordinary? When we're married will he think of me?

Chapter Twenty-six

In his experiments, Leggate had been studying putrefaction and the growth of moulds on meat. He had bought from the market a dog, and offered it pieces of decaying meat, thinking that even if the moulds that grew on them were not the causes of cholera, they might cause other diseases such as summer diarrhoea. The dog swallowed the meat greedily, without ill effect. He had to pay a second time for the man who had sold it to take the animal back.

If I can't convince myself of causal links, he thought, how can I convince others?

He made a serendipitous discovery. At dinner he had taken some hot stewed beef prepared by Mrs Huggins, and put it in a jar. He took the jar to the garden shed, cut the meat into three pieces, returned one piece to the closed jar, put one piece in another jar under muslin to keep off the flies, and put one piece in a third jar left open. He found, as he now observed regularly, that putrefaction occurred more slowly on the beef left in the closed jar than on the pieces left open or under muslin. But most surprisingly, after dinner a few days later he realized that all three cooked pieces had putrefied far more slowly than corresponding pieces of the same meat obtained before Mrs Huggins had cooked it.

Cooking must kill the planticles and their seeds, he thought. Why didn't I think of this before? Not much of a discovery, it seems so obvious. Or now it does, at least. Of course that's what cooking must do: it helps us keep free of the planticles. I'll have to start my experiments again, with only cooked food.

Leggate stared at the ceiling. Even if I solve it, what connection to cholera? he thought. Or with disease of any kind ... I'll go and discuss it again with Halliwell.

He went to lie on his couch. The first article, he thought. Halliwell's right, I must explain how the mesenteric blood vessels become permeable, and how the poison is ingested along with food. The second article will show that the poisons are really microscopic seeds of planticles, probably transported by flies. But with the effect of cooking, all the experiments with flies need redoing.

At Halliwell's that evening, Leggate said: "You really think I should send one article first and keep the flies for a separate one? I've shown that putting food under muslin to keep off the flies does retard putrefaction or it does with some foods, so that points to flies . . . but the results are not all that clear. I've got to start my experiments again."

"Why not just send off the article with the alimentary theory," said Halliwell. "If you're unsure, ask Newcombe what journal would be best."

"That's not what you really think."

"It certainly is."

"There is something you are not saying."

"I don't know about a second article, unless you've got strong new results."

"They're only suggestive at this stage."

"We've talked about it before, laughed about it—that incident in the town during the last cholera epidemic when Ayer and that other chap, I can't remember his name? When they almost came to blows."

"Ayer and Anderson."

The incident had been an argument between two doctors during the 1832 epidemic. With minimal prompting it was still gleefully recalled by the townspeople. Ayer was boasting of cures he had made by means of opium and calomel, and Anderson accused him of quackery. One of them pulled the other's nose, right there in the Medical Society. There was a rumour of a challenge to a duel. It was all over the newspaper. What better than to hear of doctors, who were suspected of spreading the disease to line their own pockets, behaving like brawlers at a tavern?

"I'm not going to pull anyone's nose," said Leggate. "I shall present a scientific idea, not quackery."

"But you would mind if your idea were laughed at?"

"I'd mind, very much."

"Every medical man has an opinion. From what you've told me, you doctors are especially keen to see what's wrong with other men's ideas. You told

me of the one in Edinburgh who tried a new treatment for cholera of which
you approved, but was ridiculed by other physicians."

"That was Thomas Latta. He showed that injecting copious quantities
of saline directly into the veins of affected persons could restore them," said
Leggate. "Well, some patients, at least, just as I'd suppose if my theory were
correct. I've been wondering if I shouldn't try practising his method on some
animals."

"Don't let yourself be diverted," said Halliwell. "Send off the first article.
Even though it may be controversial, you'll be presenting facts, and from them
you draw an inference. The worst anyone could do is disagree—but flies!
They're so specific, too easy to make fun of. Every link in the argument must
be in place."

"It could be the answer that everyone's missed." Leggate had wanted his
conclusion applauded. His voice was thick with disappointment at his friend's
words.

"I hate to say this," Leggate continued. "What if someone else publishes
first?" Then, after a pause, "I could save people's lives by the harmless rec-
ommendation of upturned bowls to cover food. Isn't that my duty?"

Leggate paused. "Did you know," he said, "cooking kills the seeds? Everyone thinks cooking makes food tender, but its real benefit is to reduce
disease."

"Well, there's another discovery. You'll have a great deal to publish, do it
bit by bit, as you become sure of each piece. First publish your article on the
alimentary theory, then one on your maps of the deaths from 1832. Say: 'Look,
here is the pattern.' Those are facts, nobody would argue. They would make
you known."

Halliwell paused. "If it's like selling pumps, you must be known and
respected. No good having the best pumps in the world if nobody's heard
of you. When you are known, you can be more speculative. Get Newcombe
to introduce you to the Medical Society."

"I'm going to do that, my maps are almost finished."

"The maps would mark you as a very sound fellow."

"I have only two more outlying parishes to visit. So . . . I'll write a short
article on just the alimentary theory—my first was too long."

"That's good."

"My article on seeds spread by flies was also too long. I shall abstract it. Most of the seeds must be harmless, though summer diarrhoea shows that some cause harm. At the end I'll mention that the same mode of propagation is most likely the cause of cholera."

Though Leggate longed to be recognized for a great discovery, the idea of ridicule at any misstep could flood his mind.

I will not be called a quack, he thought. I shall publish something sound, or not at all.

Later that week, Leggate had arranged that Newcombe would come over in the early afternoon and look at his microscopic specimens. "Do you see the idea of the little seeds?" said Leggate. "What do you think?"

"I am not sure I understand all the steps, but you wish to test your theory?"

"If you could look at my specimens. First I cook the food, that kills the seeds. Some pieces are put in jars and left open, some into jars that are covered with muslin to keep off the flies, some jars are closed."

"You want me to see if there are differences?"

"Tiny seeds of moulds get onto the food, some from the air, some from flies walking on it. I'm just concentrating on the flies, to see whether any of these moulds come from the seeds they carry on their feet."

Leggate took Newcombe to his study. "The crucial comparison is between foods in jars left open, and those in jars covered with muslin to keep off the flies."

For an hour they took turns at peering down the microscope.

"There are some differences," said Newcombe. "But, as you say, you are not so expert in making specimens. When I did my microscopic work, I found it more difficult than I thought. Your microscope is not so good either, and neither is the light."

Leggate smarted at the criticism. He felt like snatching his specimens away, to do it all himself. From the older man he managed to tolerate it. He said: "But you could show me?"

"My experience was in histology, where the principles are different. Let's take these slides back to my house. I'll see if it's better with my microscope, and I'll lend you a couple of books."

Leggate thought: I've devoted myself too much to cholera, and to note taking. I should have concentrated on microscopy.

To break the silence he said: "In Edinburgh an assistant made slides for me from the intestine, so that I could compare them with sections taken from the intestines of those dying from cholera, when it returns."

"In histology I could be more helpful."

They went to Newcombe's house, where his Ross microscope was taken from its case. Newcombe put out a book of the drawings he had made from his own slides.

"For me to look at?" asked Leggate.

"Please."

As Newcombe worked, Leggate leafed through the drawings, to see how they must be presented for publication.

After fifteen minutes of peering down the microscope, Newcombe said, "This isn't much better. We must go about it differently. You've been spreading the moulds onto the slide with a knife, and this makes the specimens far too thick. Try scraping from each piece of food just a small bit of mould. Put it into a tube of saline and mix it, then pipette a drop onto a slide, and put a cover slip on it, that will allow us to see the individual moulds, or whatever they are. Tell you what. I'll think about the problem for a couple of days, while you look at some books on microscopy."

Leggate mused as he walked home: Newcombe achieved a clarity I could never approach. All my mishmash slides. Each line of research needs apparatus, chemical reagents, a laboratory. If there were a medical school or a college of physicians, a teaching hospital... they'd have facilities, assistants. They become skilled in particular preparations, as I did for Magendie... none of that here. Perhaps someone with the right facilities, with an interest in moulds... So there are some differences between muslin-covered foods and the uncovered, but nothing clear. Or the effect of cooking... that really is a finding, but surely someone must have made that before.

Chapter Twenty-seven

It was five o'clock when Leggate reached home after leaving Newcombe's. He left the books he had borrowed, and walked directly to the Brookses' house.

In Marian's music room, after they had taken some tea, Marian said, "My mother, you know, is . . . she is . . . I couldn't have a better mother. But my father. I should tell you, he doesn't always get on with people, and he's not always easy for other people to get on with either."

"I've observed," said Leggate.

"I worry that in marrying you I may be wanting to escape from him. I do love you, but I've lived in fear of him, somewhat at least . . . I've shut myself in my room to escape."

Noticing the earnestness of her tone, Leggate began to feel anxious as he listened.

"I don't want to exchange that for servitude to a husband," continued Marian. "No, sorry, I didn't mean that, of course you would not enslave me, but . . . I don't know exactly how to say it." Leggate looked at her, thinking how odd it was for her to sound inarticulate.

"I've very few examples before me," she said. "My mother and father—she tolerates him but they're not companions, and there's Caroline and her husband."

Marian paused, "I've written a few articles, for magazines—magazines for women," she said.

Surprised by this apparent change in topic, he said, "You didn't say."

"My first was on friendship. Would you read it? It contains ideas that are very close to me."

"I'd like to read it, very much."

"I'll get it for you," she said. She went to a cupboard, and handed him the magazine.

"The reason I'm talking about it," she said, "is that my principal idea is that marriage should be between people who are complete in themselves, but who join in friendship. But it's become clearer to me that unseen forces are at work. I worry we might not succeed. The examples I know best aren't so reassuring."

"Your friend Caroline and her husband."

"She says they're not friends in the way that she and I are. She says there's a dozen things a week she tells me that she couldn't say to him."

Marian paused, and looked worried. "I wouldn't want that. I would want us always to be able to say to each other whatever comes into our minds."

At home that evening, after he had left Marian, Leggate turned to his reading. For a few minutes when Marian had spoken so earnestly that evening, he had become alarmed that she was about to express reservations about their engagement, but then he had realized it was just her nervousness in speaking of her publication.

Marian's article first, of course; he read it with a growing sense of excitement.

To be married to one who can think so well, he thought, and who writes so compellingly!

He rose from his table, and lay on his couch, imagining conversations with her, imagining a closeness.

Never really known such women, he thought, who read, and think, now there's this one, and her aunt too. How fine. And how she writes, how well she does it. And we're engaged!

This won't do, he said to himself. Microscopy.

Leggate knew of the unseen world of tiny creatures and plants, discovered by Leeuwenhoek nearly two hundred years previously. He knew of Schwann's work on the microscopical structure of cells in the body. Magendie used the microscope too, but another assistant had prepared his slides— training at Edinburgh had also emphasized microscopy, but Leggate's skills in that area had remained rudimentary.

He turned first to a slim octavo volume: *Thoughts on animalcules; or, a glimpse of the invisible world revealed by the microscope,* by Gideon Mantell, LL.D. F.R.S. It had, at the end, a collection of fine coloured plates of tiny creatures collected from a pond on Clapham Common. Mantell called them animalcules, no doubt because of their movements, though from the drawings they looked for all the world like exotic plants. They were objects of beauty, drawn with great skill from a camera lucida.

Leggate started to read "... cells the basis of all life."

Yes, of course, he thought. Schwann has shown that. But here, what's this?

... if recent microscopical discoveries are calculated to alarm the timid, by showing what slight causes may lay the foundation of fatal diseases.

Leggate felt a lurch in his stomach, his heart pounded, as if in sudden danger. Is this a stray remark? he thought. Does this man know something?

Now he read with alarm. His own priority! Had he been anticipated? Was this Mantell to take everything away? Monads, vibrios, stentors, polygastria, rotifera, the flower-shaped animalcules ... hundreds of species of micro-scopical animal and plant. Mantell described them. Every one present in unseen millions. Leggate came upon this:

The atmosphere, which is always charged with infinitesimal particles of mat-ter, both organic and inorganic, wafts in every breeze immense quantities of the sporules of plants and the ova of animalcules.

Leggate's mind, like a steam locomotive in rapid motion, sped forward. Tiny organisms don't need flies to spread, he thought. They are so minute that the air is enough. Light as the air itself. I was thinking of a few seeds, or a few hundreds, but Mantell says millions, multiplying incessantly, mil-lions upon millions. They're everywhere ... I had not made the connection.

"Dunce. Dunce!" he said out loud.

Miasma is right, he thought. Just that we didn't know what was in the air, but it's tiny seeds and ova. I should've known from the cooked meat that didn't putrefy when kept in a jar. I wasn't thinking, I was too much caught up with flies. When I thought first of the specks in the sunlight, I thought they were specks of dust, some with poison on them. Then I thought of flies ... how did it go? Then only after that did I think about planticles. I'd not thought that the planticles, or their seeds rather, are themselves the specks wafting in the air. They're everywhere, in our lungs, in our food... Here in print, written in matter-of-fact tone... If some particular species makes the poison... that would be it. Wouldn't need flies. Could just waft off the surface of the filth. Mantell has realized this? Or is he more interested in the delicate forms of these polyps?

Leggate rose from his chair, heart pounding, went down to the kitchen to find a piece of cheese.

"Aren't you going to eat your dinner, sir, I called and you said you was coming."

"I'm sorry, Mrs Huggins, I will eat. I'll take it upstairs, thank you."

Back in his study, he took a bite from time to time, being careful to keep the book from being stained, as he neared its end. Mantell was drawing towards his final conclusions:

> ... particular species, too minute for the most powerful microscope to descry, may suddenly swarm in the air or in the waters and penetrating the internal vessels and organs, assert an injurious influence of a specific character on the lining membranes, and fluids, of the human frame. And from this inscrutable agency may, possibly, originate the Cholera, Influenza, and other epidemic diseases.

Leggate's eyes failed to focus. Refocusing, he read the passage again. He sat back.

He sighed. All these years, he thought. Now this Mantell with pond water from Clapham Common has it off pat—cholera, animalcules, sporules, germs of animals and plants, injurious influence on the lining membranes and fluids. My theory exactly.

He turned the pages. Are there more details, specifics? he thought. The most important problem in all medicine thrown off in a sentence.

He went out to walk, in order to ease his thoughts.

All there, he thought, as he marched the streets. And in those other books too? Why have I not kept up with microscopy. Dunce! It's the microscope that will solve it all.

He returned to his house, looked through the three other books he had brought back from Newcombe. Two were on making preparations, and one was no real good at all. None of them enlarged on what Mantell had said. At two o'clock he went to bed. He could not sleep, his body seemed only to occupy awkward positions, none was comfortable.

As if his eyes were pinned wide open, thoughts raced. So it's Mantell who has priority, he kept saying to himself. My idea of minute seeds … I wrote out two copies, I was to do a third … but that was conceived not much more than a week ago. He calls them sporules, published last year. All my work, utterly and completely wasted.

Leggate rose from his bed, lit a candle, and went to his study. He lay on his couch and remembered his dream of the dandelion parasols. Like parasols but of infinitesimal scale, he thought. The dream was a mere fancy, but Mantell gives facts, sporules and ova everywhere, swarms of something evil. This is it—sporules of one species cause cholera. Other species cause other diseases, typhoid, influenza and so forth, every one different. Yet other species are harmless. Just like the larger plants, some beneficial, some poisonous, and one that comes from India has the special property of rooting in the mucous membranes of the intestine, penetrating the capillaries of the blood vessels, to make it a sieve.

He got up, paced about his room. When I believed that the poison might be in the air, he thought—it couldn't explain how it came so slowly from India. Let me reconstruct: that was why I thought the flies were more probable, and because no one seemed to have thought of flies, so I decided to concentrate on the flies, but if millions of sporules are wafted in the air, the disease could travel slowly because stronger storms might scatter them so they get too dispersed, only when there is a very light breeze do large numbers of them get from one midden heap to another. This needs more thought …

He went back to bed. At last he dozed.

At five o'clock, he awoke, still agitated by Mantell's book. He looked over it again, page by page. There are no more details, he thought, just descriptions of all these species of minute organisms, and intricate drawings. Why is this not more widely known? When does Newcombe get up? Can't call before eight. What if he's left the house already? Very well, a quarter to eight.

He started to abstract relevant passages of Mantell's book. At a half past seven he was at Newcombe's.

"No," said Newcombe, "you shouldn't be deterred. Every medical man has an opinion about every blessed thing. That's what we're paid for. A hundred doctors, a hundred opinions—you know that better than anyone. If you write a book, what's easier than to jot down a sentence about sporules swarming in the air to solve all the problems of disease. By chance one or two opinions hit a truth, just as you'll sometimes be dealt a complete set of court cards at whist. Most opinions are harmless: 'Cover your food with a bowl.' Doesn't hurt anyone. See what I mean?"

Leggate was startled. He looked to see Newcombe smiling. From someone else, Leggate would have been deeply wounded, but Newcombe's eyes were gentle, and Leggate's concerns were now elsewhere.

I'm being too earnest, he thought.

"Not many opinions are based on anything," continued Newcombe. "If this Mantell had something substantial you can be sure he would have written it. What you're doing in your research is a thousand times more difficult. Not to speculate but to give substance."

"I should persevere?"

"Even harder—it means that others would be disposed to accept your conclusions, Mantell, for instance."

"I scarcely have a conclusion yet... Halliwell said the same thing, my idea is just an idea. This book makes my experiments on flies totally unnecessary. I had thought the mode of propagation was the key, but if tiny sporules swarm invisibly in millions, in times of epidemic a foreign species could come from India, a poisonous species."

"Why not write to Mantell? Didn't he give his address... somewhere in London? He'd only do that if he wanted communication. Or if you think it's impolite to write to his home, I noticed he's a Fellow of the Royal Society, you could write to him there."

It was as if a jumble of steel pieces—nuts, bolts, pistons, cogs, axles, wheels—scattered on the floor, had assembled themselves into a machine: a fully functioning conception to account for all the facts of cholera. Just one piece was missing before the machine could be started into motion, and run perfectly.

If I could supply that piece, Leggate thought. This is the question—how does an epidemic spread so slowly? Work out why it takes years to come from India, not just a few days or weeks on the wind. Solve that and it's complete; sporules of a species of minute planticles, light as air, swarm from a swamp in India, then to a midden heap, perhaps they have to grow there for a while, before they swarm again and alight on food, to be eaten by unwitting people in whom they act as a vegetable toxin.

Even without the missing piece, the concept had the feel of truth—grand and satisfying. Leggate felt moved by the perfection of the answer.

At the same time, he despaired. If the essential idea is Mantell's, he thought, what could be my contribution? Or if, as Newcombe says, Mantell's idea is mere speculation, what would provide the proof? When the cholera returns, someone must recognize the foreign species of microscopic plant. But if hundreds of species of minute plants are everywhere, how can I, with second-hand microscope, fourth-rate skills, no laboratory, no assistant, no chemical agents, how could I ever succeed in identifying the species that causes the disease?

When, that evening, Leggate once more reached the sanctuary of Marian's music room, he said, "I've brought your magazine back, I'm full of admiration. How well you write, I wish I could write a quarter so well."

"Thank you," she said. "But the ideas?"

"I felt inspired by them, your idea of friendship, of equals."

"You won't demand that I obey?"

"So long as you don't demand that I do."

He saw her glance, grateful for this reply.

"So in the marriage service, at that moment, I shall say silently to myself and to God, 'but omitting that part.'"

"A perfect plan, we'll seal it with a kiss . . ."

"I'm not sure what my contribution to curing cholera should be," said Leggate, after some few more kisses. "I've just read someone's book that makes things fit into place."

"How exciting."

"Yes. No. Not exactly—he, a man called Mantell, has anticipated me. Not shown how it happens, but part of the idea would be his."

"Tell me."

"It's as I said. Disease may be caused by tiny seeds of plants, so small as to be invisible, he calls them sporules, little spores. I'd imagined a few seeds here and there. But Mantell says they're present in millions on millions, everywhere. Light as air, they swarm on the slightest breeze."

"Millions of them?"

"There's a whole kingdom of plants and animals too small to see. I brought the book for you to see." Leggate fetched the book that he had placed on a side table.

"See these." He showed Marian the delicately coloured plates of exotic flowerlike animalcules, each plate a six-inch-diameter circle on a shiny white page. "In reality this circle is smaller than a pinprick, so even in a drop of water there are hundreds of thousands of these living things."

Marian turned over the pages, gazing intently. "These are really true?" she said. "They're exquisite, like hothouse plants. So small they can be seen only with an instrument?"

"A microscope. These drawings are especially beautiful, but things like these have been known of for nearly two hundred years. The new idea is that some are poisonous. They float in the air, by their millions, then cause disease in us."

Marian placed the book on her lap, and looked at him directly, so that he felt she was following.

"I think there's one deadly species that grows first in India," he said. "Then a slight breeze carries its seeds to places nearby. They find stepping stones on the midden heaps of towns, and they get onto ships, and soon they will be here again.

"When food is eaten, that the seeds have fallen on, they find fertile soil in the human intestine," he continued, "and that's where they grow and multiply even more."

"Is that all in here?"

"No, in the book there is only the idea of sporules light as air, the idea of different species—like those that are illustrated—and the idea that some might cause disease. My contribution would be that one distinct species comes from India, and that we catch the cholera if its sporules get onto our food. Some of my research, I think, supports this."

"And this is what you will write?"

"It's not complete."

"If others are writing similar things, you must hurry. You've told me that it's important to publish first."

Whether it was prompted by Marian's tone of voice, or by an echo of his own injunctions to himself to write when he had not been able to, Leggate felt a surge of annoyance. Her words seemed to chide him. He did not reply.

"If they come in the air, all these little sporules, you should tell people to cover their food. It is a simple precaution," she said.

"Yes . . ."

Leggate's desire to withdraw from the conversation became yet more irresistible.

"You should write a letter to *The Examiner*, to recommend it, if the cholera comes again. I can speak to Warrinder, to make sure it's prominent."

"It's not ready yet. Please play something for me."

Leggate spoke evenly, but the mood between them was lost. He felt upset at her suggesting, in an offhand remark, what had taken him painstaking years to reach: the idea of covering food. And he could not bear her offering help in getting a letter published.

Chapter Twenty-eight

During the months of her engagement, Marian exuded a spontaneous joy. She had slight personal experience of the religion of love, indeed she had felt disdainful towards it, but she was converted.

Her diaries, before the engagement, caught the growing excitement.

He understands the importance of Atmosphere in the playing of Chopin. Perhaps I need an audience of only one, the right one. We will make our home together, he can see his patients, write his articles, I shall help him in his researches, encourage him to make that Discovery. I never felt so alive as in his company. That longing, I did not know it was to be filled by this. I had become sceptical of love between man and woman. As a general rule scepticism is no doubt correct, but in this instance—how could I be so fortunate? He is capable of what seems so rare in men's attitude to women. He is capable of Friendship.

Only on three occasions during her engagement was there mention of breaks in her mood. Surprised by the first of these she spoke to Caroline.

"Did you feel apprehensive before your marriage day?"

"Frustrated, I would say. I longed to set up together."

"You were never nervous about the changes?"

"I think I was, if I remember."

"Last night I had a fit of apprehension and trembling. The reason wasn't distinct, but connected to my engagement. As if by getting married something bad would happen."

"Trembling?"

"Quite severe. I have never felt so ... so ... a sense of dread, awful."

"I am sure Dr Leggate is reliable—not the sort to coerce." Caroline looked alarmed at Marian's description of her symptoms. "I've not heard of this, but he's a doctor, you should ask him."

"No, that's not it. I believe he's capable of friendship, and he's said he wouldn't want me to obey, though when we're married he'd have the right to demand obedience."

"If you're trembling, you may have caught a chill, you should look after it, be in bed."

"It's not that, I'm not ill, but I see my mother and my father . . . I don't know. Is it my duty to give over everything of myself, seek only to please him? I'm frightened that I can't stop myself wanting to give everything of myself away, just to please him, but everything I know says that's how one can lose oneself. Is that why I feel frightened?"

"Women do worry before marriage," said Caroline, evidently detaching herself from the idea of disease.

"Just a state of nervous irritability?"

"Marriage is a new state. It's natural to be apprehensive with so great a change."

Marian did not speak further to Caroline of these matters. To Leggate she did not mention them at all.

For the most part, Marian wrote in her diary as a young woman in love.

Yesterday, when we were together, I experienced with him that sense of Transcendence that I know sometimes when playing Chopin, as if I am outside myself, connected to every thing, and to every body, yet more present in myself than ever. I believe he felt it too, for he said he would give the whole world for me, to make me happy.

"Yours to command," he said. I believe he meant it, and I felt touched by it, but I also found it disturbing, and wondered if he hadn't fully understood my article. I had to explain again that I was not going to do any commanding, that we must both strive for something more equal, for Friendship. "Command" may just have been a metaphor; he agreed readily enough. I think he does understand.

But there were two more episodes of apprehension, with reflections in her diary.

Should I come to depend on him? I shall not be dependent, must not be— are these trembling fits intimations? Is that what it would be like, that shaking, an awful dissolution of self?

Would anyone be better than he? Or should I not marry? Like my aunt?

Chapter Twenty-nine

When at last the trial of Augustus Biggs began, Marian, like most in Middlethorpe, eagerly read the news in *The Examiner*. The trial offered Warrinder the opportunity to extend his literary skills from descriptions of detection to the clashes of the courtroom. Reports of the trial were printed *in extenso*, accompanied by Warrinder's commentary.

In his report of the trial's beginning, Warrinder told of how Biggs looked for all the world like a gentleman "whom you might see coming out of Trent's Bank, or sitting in the council chamber of the Town Hall."

When it came to evidence about Marie Watson's manner of death, Warrinder continued minutely to describe the events.

> Before the main case for the prosecution was to be presented, the Judge ordered first that the evidence from the Coroner's Court be offered directly to him and to the jury. He declared that Dr Tuff must first repeat his evidence, and then be questioned by Counsel for the Defence, and then by Counsel for the Prosecution.
>
> To the stand came Dr Tuff, advanced in years, bald of pate, and stooped of figure. He reiterated at otiose length the observations he had given at the inquest.
>
> Dr Tuff: "I examined with a glass of high magnification, the internal organs of the deceased. There was a brownish coloured bilious-looking fluid which surrounded the intestine. The membranes were opaque, and there were some adhesions of the mesentery which prevented facile examination of the length of the intestinal tract. But I persevered, and can report that there was no indication of ulceration, or perforation, or of haemorrhage, or of other disease. There were odours, however,

of especial significance: they were not of opium, or of prussic acid—which has the distinct smell of almonds, and is always distinguishable. But there were traces among the partially digested remains of a substantial meal which included mutton chops—more than traces, a significant quantity, of strychnine, in the lumen of the stomach..."

Counsel for the Defence: "So what did you conclude?"

Dr Tuff: "That strychnine poisoning was the cause of death, and although we have no knowledge of the circumstance under which the poison entered the body, we can conclude that strychnine was administered, and that this rather than the dagger was the immediate cause of death. The dagger caused no sputtering of blood around its place of entry. Hence one must conclude that the blood was no longer in motion."

Counsel for the Defence: "And was this examination of the internal organs sufficient for you to reach a firm diagnosis?"

Dr Tuff: "It was, sir."

Counsel for the Defence: "Thank you, Dr Tuff. Thank you, m'lord. I have no further questions for this witness."

Counsel for the Prosecution: "I am satisfied, m'lord. I have no questions at this juncture."

Counsel for the Defence, up from London, then asked if he might call a witness whom, like the man he was defending, he had also brought in from the Capital, presumably believing that men gathered from closer at hand would not be sufficiently expert. He told the Judge that this witness was not there to give evidence on the murder as such, but he wished this very great authority to offer an opinion on the evidence of Dr Tuff and the verdict recorded by the Coroner's Court. "Rather like a second opinion in medical circles, m'lord."

The Judge did not demur, and neither did the Prosecution, which was surprising since the object of the Defence's manoeuvre to attempt to cast doubt on Dr Tuff's evidence must have been plain to all. So the expert, the famous forensic toxicologist and physiologist Professor Wardle, was conducted to the stand.

Counsel for the Defence told the Court the Professor had written a well-known book on poisons. Responding to Counsel's questioning he described strychnine as a highly poisonous, bitter-tasting vegetable alkaloid. He gave a learned account of its derivation from the seeds of *Nux Vomica*, a tree of the *Loganiaceae* family. He described its solubility, its crystalline structure, and the formation of compounds with other chemicals. It was an impressive recitation.

Counsel for the Defence now moved into his stride. He reminded Professor Wardle that Watson had been seen by upwards of two hundred people, stretched out with limbs extended.

Counsel for the Defence: "Is this the posture of a sufferer from strychnine poisoning?"

Prof. Wardle: "It is utterly inconsistent with strychnine. Strychnine produces convulsions, so that the body would be drawn taut like a bow. It's an unmistakable posture."

Audible consternation in the court.

Prof. Wardle (continuing): "Furthermore, Dr Tuff's assertion that examination of internal organs is sufficient in such cases is false. A diagnosis of strychnine poisoning cannot be made without examining the brain and spinal marrow. His conclusion was too hastily reached."

General sensation in the court.

> As an observer in the court, this correspondent feels
> bound to say that when eminent authorities contradict
> each other so flatly, medical evidence begins to fall under
> the shadow of fallibility.

"May I see that?" William had come into the room. Marian, becoming irritated with Warrinder's recitations of the trial, and not wishing to offer herself as target for one of her brother's gibes about following the writer word by word, gave him the paper.

"It's not very interesting," she said. "Just Warrinder blowing his horn."

"Something at least in Mudpuddle Town," said William.

The court rose before Counsel for the Prosecution was able to cross-examine Professor Wardle. The trial had started in the afternoon of the day before the newspaper had to go to the press—a fortunate occurrence, since the newspaper, suspended on this moment of irreconcilable difference between Dr Tuff and Professor Wardle, could warn its readers that they should not miss the succeeding instalment in the next week's issue. When Marian read the next issue, she saw that readers were not to be disappointed.

> TRIAL CONTINUES OF AUGUSTUS BIGGS
> Last week we gave you all the details available up to the
> moment of going to the press of the trial of the mur-
> derer of Marie Watson.
>
> As readers will recall, Dr Tuff reiterated his evidence,
> given at the Coroner's inquest, that Marie Watson had
> died from taking the poison strychnine, of which he
> claimed to find traces in the dead woman's intestine.
> Counsel for the Defence called Professor Wardle, an
> authority in toxicology, who said that the posture in
> which Marie Watson was found was inconsistent with
> poisoning by strychnine, and moreover, that it was
> impossible to detect strychnine at post-mortem, since it
> left no traces in the alimentary canal.

Now came the turn of Counsel for the Prosecution to cross-examine the Professor from London. The Prosecutor is a man from these parts. His eyebrows were raised beneath a wrinkled brow, and his expression gave him an appearance to daunt any who might be puffed up with their own importance.

First Counsel for the Prosecution questioned Professor Wardle minutely about his credentials. It emerged that this great authority on toxicology spent his time among specimens, glass retorts, and laboratory animals. He was not so much a pathologist as a physiologist who engaged in vivisection!

Then came the hammer blow.

Counsel for the Prosecution: "Have you, Professor, personally had experience of a human case of strychnine poisoning?"

Professor Wardle: "It is not necessary in such matters. What is important are the physiological, the pharmacological, the toxicological properties."

Counsel for the Prosecution: "Be so good as to answer the question directly. Personal experience of a case of strychnine poisoning in a man or woman, yes or no?"

Professor Wardle: "No, but this is not relevant . . ."

From the uproar that was heard it was clear to all that the public gallery did not agree. Neither did Counsel for the Prosecution, who continued at length to show that as well as engaging in the repellent practice of vivisection, the experience of the so-called authority in practical cases of poisoning was negligible.

When Marian turned the page of the newspaper, she found the editorial which she expected, written by Warrinder.

In cases such as that brought against Augustus Biggs, we are apt to pay attention to medical authorities, but should we do so? Listening to one such authority, it is difficult to avoid being persuaded, but put two together, as has occurred in this case, then what do we find? The dogmatical conclusions of the one dissolve in the contradictions offered by the second. Then this second himself is found to be no better than the one he contradicts. What can possibly result but that the value of so-called scientific evidence in such cases will sensibly decline? We did not need Dr Tuff or Professor Wardle to tell us that Marie Watson was dead, or that her death had come about through wilful murder. This was obvious to the police, and indeed to every onlooker on the day she was found. But once call in the medical authorities, and all becomes obscure. One expert is shown to blunder, and then another. How can the public know that manufactured in the laboratories of such medical men are not to be found a hundred such blunders? At best the evidence of medical science wastes the time of the court, at worst it may be as murderous as the deed it professes to discover.

For the sake of English justice, one lesson of this case must be that we need merely concentrate on facts that are obvious to every right-minded man. To allow ourselves to be diverted along the path of medical authority will ensure that modern toxicology will eclipse medieval witchcraft in the atrocity of its superstition.

Chapter Thirty

Despite the urgings of his friend and of his betrothed, Leggate wrote neither to Mantell nor to *The Examiner*. He determined instead to finish his map of the deaths of 1832.

After a call on the north-west side of town, he rode to the village of Filby, a mile or so farther out, left his horse at an inn, and began to walk. As usual he took with him his notebook, his Ordnance Survey map, his compass for map-making. On such excursions he would scramble through hedges, ruin his boots crossing muddy patches where cattle stood near gates, and return home with trousers covered in burrs.

Now, as he walked, he thought dejectedly, I shall finish at least, I'll write about the pattern of deaths for *The Medical Gazette*, or perhaps just a small pamphlet for historians tracing the disease in Yorkshire. If anyone should ask, "What was the susceptibility of labouring classes?" it would show that it was not uniform in different parts . . . Not much compared with sporules swarming into the air. But it will be something . . . Then I'll give up research. Too many others, Fellows of the Royal Society, with assistants to make microscopic preparations, access to a camera lucida. It's beyond my powers . . . Research used to occupy me, but it was a fancy, carried over from Magendie who had chemicals galore, an instrument-maker, a glass-blower, apparatus, syringes, everything necessary. With new responsibilities, marriage . . . There'll no doubt be children. I should take more patients, there'll be new expenses.

He came to a farmhouse, knocked at the door. "I'm Dr Leggate," he announced to the woman who opened it, probably not much older than himself, a world away in the gnarls of her face. "Do you keep the farm here? I'm compiling an account of the cholera that struck fifteen years ago, to help prevent it occurring again. Could you tell me if anyone in your family was affected, or anyone hereabouts?"

Leggate's friendly manner reassured the woman. "Come in, doctor. I'll make you some tea. You say anyone affected? It was almost everyone here.

My husband, he was the first who got it, then his father who lived with us. Both of them." She led him into the kitchen.

"Our firstborn caught it too, though she survived, thank God. And a boy who lived in, he died. My father had to come in, to help me keep the farm going. I'm married again now, become Mrs Ericson. My husband and I work the farm."

"It was a terrible time. Do you mind if I write a note about it."

"Please do, doctor. Sit down for a moment."

While the woman busied herself, Leggate looked around the kitchen with its homely furniture, the large wooden dresser with cups and crockery, the pans in the fireplace.

She kept her back to him, for a tear still would spring at the recollections that he had asked for. He wrote in his notebook, gently asked more questions about who had caught the disease, when, what had happened. The woman answered, back still turned, fussing at teacups.

"I am sorry to bring back these memories. I wonder, what about your neighbours? Do you know them? Did any of them go down with the disease?"

"It's a strange thing," she turned at last to face him, her face puffy. "Everyone talked about it at the time, wondered if it weren't some curse. Come outside and I'll show you."

She led Leggate out of the house, walked across the uneven yard spattered with straw and the dung of animals, a cloud of flies rising. She shooed off clucking chickens, pointed across a stone wall.

"Over there towards the town, see." Middlethorpe was in the distance, its smoke pall smeared out to the left.

"See the line of houses," said Mrs Ericson. "This one and four others, every one was visited. One family, the mother and father both. You can see their farm, the one after our neighbour, see their barn with the roof half off, never repaired after the storm—the house is still empty."

"And your neighbours over there?" said Leggate, cast down by the sombre sight, as he pointed now eastwards across flat farmland. "What about them?"

"That's not the direction it came. As if the Devil drew a line a mile long and said, 'Those people there. Those five houses, they're for it.' You go over to the church, St Stephen's, that's our parish church, you'll see the graves, all the same year, 1832."

"What about that farm over there?" asked Leggate now facing into the wind. "Do you know those people?"

"That's Ledbury Farm. They've thrived. Always have. It's their own. And the estate beyond that belongs to Mr Sykes, you can see Ourby Grange over by that copse, see." She pointed to a large house a mile away. "Owns a lot of the land hereabouts, including this."

"So you're tenants."

The woman laughed. "Lor' bless you. What else do you think we'd be?"

Leggate drank some tea, ate some bread with a piece of pungent cheese that the woman proudly announced was made by herself. "You can get it at the market on Wednesdays, if you like it," she said. "I'm not there every week. Take a piece with you now. I take turns with my brother's wife. You can get the cheese, or eggs, or anything else."

When Leggate left, carrying a piece of cheese exchanged for a coin, he was uncertain whether to go back to the village and straight to the parish church, or to walk the line of the curse on these fields. He walked the Devil's line, a rough track ran in the right direction. He reached the first farmhouse; there were animals and geese.

No sign of people, he thought. Must be in the fields. I'll come back later.

He walked onwards towards the next farmhouse, the deserted one. The windows on the ground floor were broken. From one appeared a young tree. He went round to the yard doorway. There was no longer a door. Inside the stone-flagged kitchen there was a smell of animal excrement. A mournful table lay capsized among a pile of rubbish. A wall down which water had run was covered by an egregious mould. Ghosts of a once determined family had retreated, nature had reasserted herself. He was disinclined to penetrate farther. He turned and pushed his way through high weeds, towards a gateway on the other side of the yard. He jumped onto a stepping stone to cross a little brook, struggled along what must once have been a clear path, towards the church. There he spoke with the rector, who looked up the registers, showed him the graves, and identified the affected houses on the map.

Later, in his study, Leggate recorded his notes. On his master map, he drew the five houses in a line, accompanied by numbered circles for those who had died. He performed all this out of habit, the line of houses a quaint accident.

First to die were in the house of Mrs Ericson, he thought, formerly wife of George Beeston; she must have been young when her husband died. First Mr Beeston took ill and died, then Beeston's father died soon after. A child caught the disease too; she survived, but the young farmhand did not. Three people in all. Then two days later, people in the next house, two died. Then in another two days, deaths in the houses beyond that.

Then he saw it.

"Five farmhouses on a stream!" he said out loud.

The stream I crossed as I came from the deserted house, he thought, the seeds of the disease carried downstream from house to house ... The water. That's where the animalcules and planticles live. In the water, like the pond on Clapham Common. Not food, it's the water!

Words tumbled onto his notebook.

Found at last. More certain than the flies, or sporules swarming in air. In the parish of St Stephen's, five farms—what connects them? Not a line drawn by the Devil—it's a stream, just a stream, a tributary of the River Keel, called Ringers Beck. Five farms, all have their houses built on the beck. I did not ask. It's dark now but I shall go tomorrow, and I'll wager a hundred pounds there are no wells, but that they take their water from the stream. It meanders a bit, but by chance the farmhouses are pretty much in a line, then the stream runs through fields, till it reaches the river. In each of the five houses people died from cholera in 1832.

This clinches it. If I ask, and they all say they take their water from the stream, it is certain. What more proof is needed? My maps were the most important thing, after all.

The disease strikes, sporules are carried out in the evacuations which find their way into whatever drainage exists, in this case into the beck, which is the very source of water for others downstream. The disease grows in the intestines of any who drink the water, drawing fluid from the blood into the excreta, which then carries away the seeds to be sown anew.

He paced the room, lay on his couch, stared at the ceiling, returned to his writing table, took up his pen again:

Those attending the sick person and handling any bedding or clothes may carry the particles on their hands, then if they take food, the particles enter the mouth. This is why doctors attending the sick do not succumb, because in Sunderland and Paris, Magendie insisted we wash our hands—why did he not develop this idea? it is the key.

Leggate tried his idea on the facts he knew about the town. I don't know enough about water supplies in different parts, he thought. Tomorrow I'll check. What about the workhouse? Always an anomaly, the poorest people, but only one death. The pump in the courtyard? Iron pipe with wooden handle, if I remember, it must draw from its own well. I'll find out tomorrow.

The candles guttered low. Leggate lit two new ones, pressed them into the stumps of the old. Excitement propelled him. He turned again to his notebook, intent on catching every errant thought, not concerned whether he wrote coherently.

At last the full solution, and my alimentary theory correct, not flies, not air, but water, when all the time I was thinking of propagation by food. But it is water, the very most intimate stuff of life. Why did I not understand before? why did not others? Microscopic plants and animalcules, known for two hundred years, and people drink this water, which contains minuscule seeds that take root and grow.

First signs of cholera always copious diarrhoea, rice-water stools, fluid sieved from the blood, and vomiting. Minute plants are first taken into the gut, like so many dandelion seeds, take root there, their long roots burrow through the wall of the gut, make holes in the blood vessels, through which the fluid of the blood is sieved.

Are roots of these little plants withdrawn to make the holes? Perhaps the tiny plants dislodge, and the fluids seep into the gut as diarrhoea, then the little plants are carried away as this fluid is voided, and when this fluid gets into any supply of water, and someone drinks it, the little plants, or their seeds, or whatever they are, can reach a new victim.

Leggate rose from his chair, took one candle to the mantelpiece to inspect his clock. He wound up the chimes. Half past two. He paced the room, then lay on his couch.

His thoughts so recently sunk in the melancholy of defeat now sprang up: Discoverer of the cause of cholera.

He imagined Marian standing by his side, proud of him. He imagined the Secretary of the Royal Society:

"Tell me, Dr Leggate, how did you come upon this, which has defeated all previous enquiry?"

"I cannot say with certainty. I have no special abilities beyond making notes and maps."

He could not sleep. Before dawn he was out.

Thank God farmers rise early, he thought as he rode towards the Ericsons' farm.

It was too dark to make his way easily, but gradually a thin dawn allowed him to see the road. He reached the farm. The pale form of the woman he had spoken to on the previous day emerged from a barn, and walked towards her house with two buckets of milk.

"Why, doctor, you're up early. Did you leave something behind? Not more cheese?"

"Good morning. No, I didn't leave anything. Just a question I forgot to ask. How do you get your water? Have you a well?"

"No, the stream. It runs over gravel just over there, so it's clear. Quite good water, we take it for ourselves and the animals."

"And do you have a privy, with a cesspool?"

"What a thing to ask."

"Could I see?"

She led him to the shed, upstream a little from where the water was taken. He went round the back of the shed, to see, in the still partial light, lush weeds sprouting on the slight slope down to the stream. He noticed in the shed there stood a bucket, and through the weeds there was a worn path, less than ten yards to the stream, close enough to dip the bucket and fetch water.

"And the other houses, the ones downstream, the ones in the line towards the town that you pointed out to me, do they have wells, or would they take their water from the stream, do you think?"

"I know the next farm does. I can't say about the others."

"I can't thank you enough. I believe you have given the clue to the cause of cholera—it is carried by tiny seeds, sporules, too small for the eye to see, carried in the water. The seeds enter the water, and when people drink it they catch the disease."

The woman stared at the strange doctor making strange pronouncements.

"Well, I don't know."

"Thank you, Mrs Ericson, thank you. Goodbye. I'll come to the market, buy some more cheese. Goodbye."

A man appeared from a shed, looked questioningly at the unusual sight of a gentleman mounting a horse in his yard.

"This is a doctor," said Mrs Ericson, "asking after deaths from the cholera here, and about our water."

"I'm Dr Leggate. Good morning to you. Your wife's been most helpful. Solved a puzzle that has defeated the best minds of the age. I must go now. Thank you again. Thank you."

Leggate rode off. In an hour he had visited the other inhabited farmhouses. All took their water from the stream.

"And the deserted house? Did they have a well?" Leggate had asked.

The neighbour knew they had not. "The stream's not bad. Country's very flat here, so it's not a fast stream, and it gets low in the summer, but there's always at least a trickle," he said. "And there are pools, but it never completely dries. Less nuisance than a well."

Leggate sat on his horse, amid the wide countryside, discoverer of the cause of cholera.

Chapter Thirty-one

Still early in the morning, Leggate was at the Halliwells' eager with his news.

"What set it off was the beck," he said. "And I found cesspools that leak into it; in the summer the stream is only a trickle, but all five houses took water from it, each one below the cesspools of the others. Every one visited by the disease. It's the first definite evidence of how it spreads. If I can trace the supplies within the town as they were in 1832, I can compare with my maps. That would confirm it."

Halliwell hurried his breakfast. The two men strode through the town to the office of Mr McPherson, the Town Surveyor, who had prepared the water scheme. McPherson displayed his maps of the previous supplies, showed the plan of the new scheme.

"May I come in an hour with tracing paper to copy the supplies as they used to be?" asked Leggate.

Mr McPherson glanced at Halliwell, who smiled reassuringly.

Leggate rushed home. He called on one patient, asked his housekeeper to go round and say to another that he would come the next day. As he traced the old water supplies, then ran home to compare them to his master map, his thoughts were in a fever.

Late that afternoon, Leggate explained his discovery to Newcombe.

"You have it," said Newcombe. "I feel proud to be a friend."

"It isn't some vague emanation. I can't stand such reasoning. There's a real cause."

"You must alert the town."

"I should?"

"With Halliwell, write a letter to *The Examiner*. He's respected as an engineer, knows about water. The two of you could make a formidable case."

"Will you join us?"

"All I've done is lend a book."

That evening Leggate brought his roll of maps to Halliwell.

"Everything's here," he said. "I've made tracings of McPherson's map—to fit over my map of those who died. I've a method of changing the scale, you see, to fit exactly. Just here for instance, see this open drain, and these pump wells, the water from the drain must have seeped into them. And in this region there was a water company that drew water from the river. I used to think deaths here were because the area's a poor one, but it's not that. The river must have got contaminated. There's almost complete correspondence. Where you live, most people got water from Tafts Spring. Only two deaths in this area. Then there's the workhouse, one of my best examples. People are destitute, but only one death, because they have their own well, away from any waste or drains. Perhaps the man who died ate something on which flies had settled."

"So with the new waterworks supplying almost the whole town from the river, it will hardly be a blessing if the disease returns," said Halliwell.

"It'll be our destruction. I examined some of the new town water microscopically. It's full of particles, clearly visible, that aren't stopped by the filter beds. If sporules get from the sewers into the river, the filters won't stop them."

The two men sat up late to compose a letter to *The Examiner*; it went through several drafts, until at last they had a version with which they were satisfied.

To the Editor

Sir, it has been discovered that the spread of epidemic cholera, also known as cholera morbus, is principally by way of the water that people drink. One of us, Dr Leggate, has shown by comparing maps of the water supplies in the town with areas of greatest mortality during the cholera epidemic of 1832, that the principal cause of death in that year was the supply of water, which became contaminated with sewage that includes the evacuations of victims of the disease. When cholera is prevailing, the sewage carries seeds of tiny plants that cause it. Particulars of this discovery will be communicated to *The London Medical Gazette* in the very near future.

Our purpose in writing this letter, one of us as a doctor and the other as an engineer with experience in supplying water, is that we believe that the decision by the mayor and aldermen of Middlethorpe to provide for our needs of water from a single source, by taking it off from the River Keel during low tide, could be highly detrimental if once again the seeds of the disease get into the river. The whole town would be exposed to grave risk; moreover, we believe that the filter beds through which the water passes on its way from the river to the town may be inadequate to restrain these seeds.

There are reports of a new wave of cholera, already in Russia, which is moving towards us, hence we propose that the water from the river be supplemented by one or two other sources, and with a system of valves so that if cholera were to strike Middlethorpe again, we should be able to choose among several sources to supply the town. A new well could be sunk at Tafts Spring, where there is a plentiful underground source of pure water, such that people taking their water from this source in the epidemic of cholera in 1832 were exempted from the disease. A reservoir should be built, and a pipeline laid to the waterworks so that water could be pumped into the town's pipes from this source. Other sources would also be amenable to test, but we believe that the Tafts Spring water is of a purity that can be trusted, and that it would be prudent to have water from this alternative source available in time of emergency.

Yours faithfully,

John Leggate, Esq. M.D.

Thos. Halliwell, Esq.

Next morning Leggate walked to George Square, to the offices of the newspaper. There it was, with the lettering "The Middlethorpe Examiner" high above the second storey, showing up surprisingly tastefully on the fine building constructed in the style of Inigo Jones.

Leggate walked across the wide square, past the equestrian statue of a medieval knight. He looked round the square. All the important things, he thought: the Town Hall, banks, shipping companies, insurance, and, of course, the newspaper—all here, the centre of things hereabouts. This will give them something to think about. He went up the steps into *The Examiner's* building to deliver his letter.

Chapter Thirty-two

"I've solved it at last," Marian heard Leggate make his eager announcement when she saw him that evening. "And I've written to the newspaper."

"But didn't have time to write to me." She was cool.

"Please forgive me." Leggate tried to win her with a smile. "Dearest."

"Do you deserve forgiveness?"

"I didn't promise I could write every day, sometimes work..."

Leggate moved towards her. "Please." He leaned to kiss her cheek. "I'm very sorry, I was..." He remained close to her.

"You were taken up with discovering, and writing to the newspaper."

She saw him so excited that she felt her indignation dissolve. I mustn't make such demands, she thought, now he's here... but that waiting, such a propensity for dependence, I mustn't let myself be like that, I mustn't, I mustn't.

She put up her hand to touch the hair at the back of his neck, stood back, made a tiny shrugging gesture.

With the atmosphere almost restored, she heard what he had been so clearly bursting to say: "I have made the discovery. I've really found it."

"That it really is spread by little seeds?"

"Yes, but they're carried in the water, that's the key."

"And you've written to the newspaper?"

"They said they thought it might not get in this week, because of this murder trial, more likely next week or the week after."

"You spoke to the proprietor?" Marian shuddered at the thought of him speaking to Warrinder.

"I handed it over, in the usual way," he said.

"Have you got it? You made a copy?"

She began to read the copy that Leggate produced. As she read, she saw him more and more uncomfortable, suspended on a thread of anticipation. She nonetheless took longer than necessary, her irritation still slowly evaporating. She considered how she might speak to him with the candour she wanted between them.

However can I tell him that I once had a proposal from Warrinder? she thought. But of course I must say that I'll speak to Warrinder—he was very cool in the note he wrote to congratulate me on my engagement. Angry, of course—after I'd told him I'd never marry. But I must make sure the letter's given prominence. Perhaps for the best—it would flatter Warrinder for me to ask something of him, I should speak to him in any case, try and make my peace.

"It's an excellent letter," she said at last. "Good to warn the town too. I can talk to Edward Warrinder. He has the editor's ear; he could make sure the letter is properly displayed."

"I wish you wouldn't. It's not something that needs arranging. It's a scientific thing. Truth speaks for itself. I have done everything necessary."

There was a long silence.

"If you prefer." Marian was relieved but puzzled. A minute ago he had been exuberant. The atmosphere between them felt awkward, with something obscurely wrong, a kind of numbness between them.

Have I betrayed myself, let him feel my awkwardness when I was thinking about Warrinder, she thought, and so he's upset that I wasn't enthusiastic enough about his discovery? I must be brighter.

"I could, you know . . . help," she repeated. "I'd like to help. I've known Warrinder a long time."

"It's kind," he said. "I think best not. It'll be printed next week."

She saw him try to reassert a positive tone. "When I write my paper for *The Medical Gazette*, and when recognition comes, won't you feel proud to be engaged to someone who's done something significant?"

"I shall, and I could help, copy it for you, I've a good hand. I can help with construction of sentences—help get it sent off."

"That's very kind, but not necessary. It's almost finished. I shall send it off soon."

Marian kept silent as she rose, went to a cupboard, sorted through some books of music.

Chapter Thirty-three

Taking up the newspaper the following week, Marian searched for Leggate's letter; there were letters about the trial of Augustus Biggs, but not his.

The case for the prosecution and the beginning of the case for the defence were, however, revisited by Warrinder before he offered his account of the trial's closing stages. Marian glanced over the columns of print. The Prosecutor had the jury on his side. He had been all contempt towards the Professor from London, kind consideration towards Millie when she had identified Biggs as the man whom Marie Watson had known, polite deference to an apothecary who said that poisons of all kinds were not difficult to obtain, incredulity towards Biggs, an engaging charm towards the jury. With each new phase he sharpened his portrait of the debaucher whose warped tastes had taken too strong a hold. Biggs had tired of Marie Watson—no doubt she had demanded marriage, or told him she was with child. In either event he had drugged her and then, in a sacrilegious demonstration borne of a deranged mind, he had taken her body to the west door of the town's principal church, and plunged a dagger into her heart.

The case offered by the Counsel for the Defence was that Mr Biggs was indeed a respectable man. Evidence was produced that he attended church on Sundays and even took notes during the sermons. Chambermaids were brought on to say that he was "a very pleasant, nice mannered sort of gentleman," always ready with a shilling. He gave evidence on his own behalf.

> Counsel for the Defence: "You say you did know Marie Watson?"
> Defendant Biggs: "I did know her, I was sorry for her, wanted to help better her condition."
> Counsel for the Defence: "And to this end, on the

evening in question, you had taken her to Cockspur's
Restaurant?"

Defendant Biggs: "I had. My interest was purely phil-
anthropical. I wanted to help her make her way up from
the gutter. I wanted to persuade her to take up some
trade, such as dressmaking, or selling perfumes, with
which I could supply her."

Counsel for the Defence: "So there might have been
some profit in this, for you?"

Defendant Biggs: "There would, but I insist that was
a secondary consideration."

Counsel for the Defence: "Please tell the Court of
Miss Watson's condition when you last saw her."

Defendant Biggs: "I left her in perfectly good health,
and I took the last train back to Leeds."

Warrinder had no difficulty in convincing his readers that the case for the
prosecution was correct. He described how the eyes of jurymen had been
fixed upon the Prosecutor as he walked back and forth, how they would
applaud with each blow he struck, and were visibly shaken at each setback
contrived during cross-examination by the Counsel for the Defence.

The report of the Prosecutor's summing-up concluded thus:

Counsel for the Prosecution: "Before you, gentlemen of
the jury, you see one who has maintained a façade of
churchgoing and shilling handouts, while at the same
time consorting with women of the kind you have seen
here in Court. Of course, when confronted, such men
do not own their real intentions. They say their interests
were innocent. Of course they do; they would do, would
they not? Were any one of us caught in some more minor
misdemeanour, would we not hastily put a good face
upon it? Of course we would. But let me tell you this,

gentlemen, those creatures of the night exist for one purpose only. Men associate with them for one purpose only. So what should a man do when faced with evidence of such an association—the only slight straw to which he might cling is to claim that, unlike other men, his purpose was to help the creature: he utters the cry 'philanthropy' as if this insubstantial tissue would conceal his more base desires.

"Whatever your feelings towards such a man, and towards such women as these from whom we have heard in this case, our task today is different. Depravity cannot be approved, but however much we disapprove, it does not deserve the ultimate penalty. We know, however, that one vice too easily leads to another. The vice of which we have here heard incontrovertible evidence lacked but one step to the most despicable of crimes. This is the step the prisoner took because he was alarmed at exposure. We cannot tolerate such men in our midst. Your duty is to find him guilty."

Counsel for the Defence repeated his implausible contention that Biggs was a respectable man caught in a web of circumstance, who had only been doing his best to assist the murdered woman.

Readers may be told immediately that the jury took very little time to reach their conclusion, so transparent was the case. Augustus Biggs was found guilty, and was sentenced, in ten days, to be taken forth to a place of execution, there to be hanged by the neck until dead.

The hanging takes place a day before the next issue of this Newspaper is due to be published, and your Correspondent will attend to give readers an account of this occasion.

On the editorial page, Warrinder summed up after the verdict.

Despite attempts to derail the proceedings by medical
authorities, the conclusion of the trial was as should be.
A plausible blackguard faces the gallows, a lesson to all
who might be tempted; crime will be swiftly discovered,
and punished to the full extent of the law.

Marian pondered. Warrinder's succeeded in getting the town behind him,
she thought. Perhaps he's right. A rift in the social fabric is repaired, ordinary life can continue. But how uncertain it all seems—so many loose ends,
nothing of ill-intent towards Marie Watson on the part of Mr Biggs, nothing to indicate she had any fear of him, nothing to show she really was poisoned, nothing to account for why she was found with a dagger through her
heart in front of the church.

Chapter Thirty-four

When Marian saw Leggate on the evening of the newspaper's report of the ending of the trial, his letter had still not been published.

"It's such an unfortunate coincidence with this trial," she said.

"I went to the newspaper offices again," said Leggate. "They said my letter had been read with interest, and space will be available soon."

"Why don't you let me speak to Warrinder?"

"I don't want you to. It's just pusillanimous of them. As if these sordid matters need to obliterate everything else."

"The hanging's next week. The paper's promised an account, but then it will be over . . . but it's all ephemera; just think of your discovery, which will be lasting."

"Medical people always have their own opinions, and they want to hold on to them, so when I write I have to be utterly compelling. But I've been thinking, now it's clear who my adversaries are."

"A kind of prosecution you think, so that you'll need a defence?"

"Perhaps that's it. Those who believe in contagion won't like the idea that the disease spreads at a distance by water; they think people must touch the victim or the victim's clothes, and the miasma people like air as the medium of communication. I go against both camps."

"Let me help. I'd be good with adversaries."

"I shall be all right. It's a matter of being very clear with my evidence."

Next week the newspaper printed Warrinder's account of the hanging.

> Never was there a more orderly or sombre crowd than those who circulated round the scaffold, looking up to see the dark outline of the gallows. A minute or two before the appointed time the prisoner appears, hands bound before him, he mounts the steps of the machine

which is moored to the masonry on either side of the prison gateway. The prisoner lowers his head so that the hangman can slip the rope over it. The prisoner raises his bound hands to put his fingers between the noose and his neck, as if he is wearing a collar that is too tight. The cap, more like a flour bag, is placed over his head, and the hangman hurries down off the machine, to pull the lever that draws the bolt of the drop. Immediately, as the body falls through the trap, the hangman leaps to grasp the thrashing legs to make sure his job is complete. He lets go, and the body, no longer convulsing, revolves and swings slightly.

On the day after the newspaper's report of the hanging, Marian and her maid, Lucy, were on an errand, and saw Warrinder in George Square. Marian had not spoken to him directly since he had sent his perfunctory note on the announcement of her engagement, and she had sent her brief reply. Lucy saw that she should move a few steps away, and went to look in a shop window.

"You think me weak and vacillating to engage myself to be married," Marian said, blurting out her immediate thought.

"It was to be expected," he said.

"For a woman to change her mind."

"To find some reason to evade a principle."

"Now that is cruel ... not the substance of friendship."

She felt his pain, but at the same time resented his stab. He acts as if he had some right! she thought.

"You are satisfied with the result of the trial?" she asked.

"You're not to tell anybody," he said, his sullen demeanour changing on the instant. "My uncle has decided to retire at the end of this month. The newspaper's done so well with this case, I'm to be editor, and have a share of the proprietorship."

"I don't see why I should keep it to myself," she said. "I must be the hundredth person you've told. My congratulations, in any event."

"Why, thank you, ma'am." He made a mock bow.

"It is an honour for one so young," she continued.

"Trouble is," said Warrinder, "Jack Peasgood's got a printing works, he's been so impressed by the success of *The Examiner* that he's started a newspaper, *The Middlethorpe and District Mail*. He claims it'll serve the whole area."

"I've seen it."

"I don't know if there's room for two newspapers, but I shall certainly make ours the best. We live in a great town, and I shall help make it greater."

She saw him soaring.

"The gallows scene," she said. "From the way you described it, it seemed to upset you."

"I've never attended a hanging before. It was appalling."

"To think that you were responsible for sending him there."

He was silent. He looked downwards. When his eyes next met hers, she saw they were those of a wounded animal.

"I hope we did right," he said. "Seeing someone put to death... He should have been sent to Australia, or something of that sort."

"You became excited with the chase," she said. "You got the town excited with you."

"I get caught up in things."

"You once asked me if I would offer my views on pieces you would write."

"I..." Warrinder shrugged his shoulders, almost apologetically.

"So the result of Marie Watson and Augustus Biggs," she continued, "and of your proprietorship is that you won't need to take account of anyone's views in what you write."

"I have to take notice of the truth."

"The truth?"

"And public sentiment, otherwise we couldn't sell our newspaper."

Next day Marian received a letter from Warrinder.

Dear Marian,

　　When we met yesterday, I had much on my mind, with the announcement of my new position, but I have to own that I felt most deeply wounded after I said how the sight of the hanging had affected me, and you said that it was I who slipped the noose round Biggs's neck. It felt almost as if you wished purposely to hurt me—I wish this were not so, but the incident revived vividly, for me, the recollection that I thought I had succeeded in putting aside, when you rebuffed my offer of marriage. I understood you then to say that you declared for spinsterhood, and intended never to marry, so that now that you are so intending, I find I revisit what we said to each other on that occasion. Something, I suppose, must have changed, for I gather that you will be married with the full ceremony of the Church, in which you will promise to obey. I can only imagine that you feel able to obey Dr Leggate, whereas you would not have been able to obey me. Forgive me for being so blunt—the matter revolves in my mind. I fear you were not honest with me, so it is a hard struggle to maintain my good opinion of you.

Please excuse the intruding nature of my letter—I would that we could remain friends. I fear that we may not, but I hope there may be something that you can say to reassure me.

Yours ever,

Edward Warrinder

Marian let the letter fall, feeling hollow emptiness in her stomach—she was rightly accused, she had been false. She picked it up again, read it over. She went quickly to her writing table to compose a reply.

Dear Edward,

I feel you most unjust to me. I am your friend—it is true I did goad you about the hanging, and that was very, very wrong of me. I don't know what impulse seized me. I am abjectly sorry, for I was cruel. Please find it in your heart to forgive me.

And as to my engagement, and to your offer to me so long ago . . .

Marian stared at the wall, not knowing what to say. What is there to say? she thought. He's hit the truth, I fobbed him off with a generality, when it was he, specifically, I refused. She continued the letter:

... you remember I said at the time that I was not suitable as a wife, only as a friend, that I was too contrary, too opinionated for a wife. That was true. If I now link myself to another, then you may think it will be he who suffers my lack of feminine softness, my cruelty—of which you recently received an example, and of which you rightly complain. Think that you have had an escape. I do not know what else to say, except that I am sorry, and that I truly hope that in your generosity, which is far larger than mine, you can forgive me, and that you can believe also that whatever it may seem, I am, and will continue to be, your very close friend.

She signed the letter without reading it over. She imagined a kind of secret bond with him, a friendship outside the conventions of marriage. She thought he would grasp the idea and allow it to heal the wound she had dealt him. She sealed the letter, ran downstairs to give it to the maid to post immediately. She rushed up once more to her music room, where she lay on her *chaise* and wept.

Chapter Thirty-five

When, the week after the newspaper's account of the hanging, Leggate's letter had still not been printed, he marched again to *The Examiner*'s imposing office in George Square. At last he was given a promise: his letter would be printed the following week.

Next week the town will know of my discovery, he thought. The excitements of the trial will be no more, and people can concentrate on important things. They will point me out: "There goes Dr Leggate." And my finding will show how both theories work. Contagion works by the sporules evacuated by one person being touched with the hands, before eating. Miasma works because there are places with poor drainage, where the probability of disease is therefore increased.

Leggate waited for a cart to pass, as it lurched through the mire on the corner.

My finding will have the widest significance, he thought. Practical solutions. Water supplies must be rethought, but Chadwick and the sanitarians have something, nuisances must be cleared and drains covered.

At home, in his study, he took fresh paper. This would be his gift to the world. He began.

TO THE EDITOR OF THE

LONDON MEDICAL GAZETTE

TWO EVIDENCES THAT CHOLERA MORBUS

PROPAGATES BY THE MEDIUM OF WATER IN

THE SUPPLY TO TOWNS

He crossed out the first word of his title, and crossed out the last letter of the second.

"Evidence," he thought, not "Evidences." More modern. Now what?— "medium of water in the supply to towns" seems hopeless. They could say I don't know because my evidence is only of one town. I'll change that later.

He underlined the last phrase of his title, then added a question mark as a reminder to replace the last phrase.

He dipped his pen, and began to write:

The means of transmission of cholera has remained, throughout the period of modern medical research, largely unknown. Although some authorities aver this fatal scourge to be spread by contagion, others opine that its effects occur by way of local miasmata.

He read what he had written. The first sentence isn't bad, he said to himself. But why "largely"? The cause has been completely unknown. "Completely" would be better, but it sounds arrogant. Perhaps cross out "largely," have no qualifying word, just "remains unknown." Yes. But the next sentence, "fatal scourge" sounds pompous, not suitable at all.

Leggate crossed through the whole sentence. He crumpled the paper, began a fresh sheet, writing the modified title. Then he crumpled that paper, and wrote on another piece, "Evidence of the propagation of cholera."

There will be argument, he thought. I must consider the adversaries.

He walked to the other end of his room, to take up a position opposing his desk. What would the contagionist say? he thought.

He paused. He could say, how do you know that someone didn't leave the first farm and go to the second? perhaps to get help when Mr Beeston became ill, they are neighbours after all, perhaps this person, perhaps Mr Beeston's father, had taken the disease, and then by shaking hands, gave it to someone in the next farm. Then someone in that farm becomes ill, then their neighbour farther downstream comes to visit, so they catch it, and in a few days the disease spreads to the whole row of farms—that's what someone could say.

He returned to his writing table, and faced now in the opposite direction, half sitting on the table.

It's not true, he thought. Too implausible—the water makes far more sense—but how do I show them? I can say that it's not just the farms, there's the close correspondence with water supplies of the town, and the workhouse. How would they explain the workhouse?

Leggate rose, opened again his window to the shouts of dockers, and the thud of deal timber unloaded from a Baltic port, saw that the ship at the staithe opposite, the *Isaac Newton*, was being winched out into the river, but then had to be winched back to await a vessel coming upriver.

It's astounding there aren't more collisions, he thought. That boy whose leg was crushed by a barrel when those ships collided on Saturday. I need quiet—in the new house, with Marian, that's what there will be. I liked this scene when I was first here, but noise is not conducive.

Leggate went to his table. He dipped his pen, made some strokes. Unsatisfied, he dried the pen on some blotting paper, took it to the window to inspect the nib with a magnifying glass, decided it had become bent, put the pen down, started looking for the new pens he had bought. Before he found them, he lay down again on his couch.

Why don't I just simply write what I know, he thought. Point 1, the cholera comes nearer each week. Point 2, its means of transmission not previously understood... not even "swarming in the air." Point 3, evidence from the deaths in the houses along the beck in 1832, a table showing the names of those who died, and their dates. No, the names aren't necessary—what are they to physicians in London? What about the microscopic plants? Should I discuss the research of Leeuwenhoek and Mantell and other microscopists? An unseen world of such things, and not all the minute creatures and plants are benign . . . No, stick to my facts, and draw the conclusion cautiously, say "one explanation would be . . ." and just say "substance," or "poisonous particles." In a concluding paragraph mention Mantell. I could quote from his work about the sporules being responsible for cholera and influenza. Other epidemics and zymotic fevers must work in the same fashion, typhoid, influenza, all the others, each one due to a distinct species of planticle. Just find its means of propagation, and we will have it.

With an article in *The Medical Gazette*, Leggate mused that his ideas would have the force of published work.

I've written a letter to the newspaper, he thought. I write notes every day. Why is this so difficult? I have new facts, the farms on the beck, and the deaths following each other by days, these are facts—all that is necessary for science. Be plain and direct, simply a matter of transcribing my notes. The contagionist's and the miasmatist's arguments can't be right—they can't explain the distribution of deaths in the town, or the way in which the workhouse was spared; I can. Was it because Halliwell helped that the letter to the newspaper was finished so quickly?

He rose, saw a piece of paper towards the back of his table, on which a previous version of the opening paragraph had been written, in fair copy. He read it over, then searched again for his pen.

I put it down somewhere, he thought. I was going to start a new one. Where are they?

He went to the window, opened it to the hubbub of the familiar scene.

Busy, busy, he thought, purposeful, wheels of commerce, why not of science? In Middlethorpe we can do science too.

The sun neared its zenith. What's she doing now? he wondered. In our new house she would be downstairs . . . I could go down and say, "I need to pause for a few minutes, collect myself."

So Leggate fell to thinking not of the difficult subject of an article for a scientific journal, but of her. He put away his article, found his pen, opened his diary.

With Marian, I have no doubt of my acceptance, I feel her love, its warming presence; she the most perfect creature, what does she love about me? My science, yes, for I believe she senses I shall achieve something, but between us she seems appreciative that I understand her aspirations, music, and that I respect her, so we shall live so that we are both respected, here in this town, perhaps later in London.

Chapter Thirty-six

The following week Leggate was excited when he opened the newspaper. His and Haliwell's letter had pride of place, first of the week's letters. He read it through, twice, carefully, saw with approval that it had been printed without mistakes. It looked very fine, exactly to the point. He bought more copies of the newspaper before going out on the day's calls.

It's only a town newspaper, he thought, but it's a start. I must have been waiting for the publication. I couldn't get properly launched with the article for *The Medical Gazette* until it was out. Now it's just a matter of writing what I've found; other contributions in *The Gazette* are not so distinguished that I can't surpass them. After all, this is the problem that fifteen years ago every great mind in medicine was longing to solve, and when the disease comes again, it'll again be the most important.

He looked forward to meeting Halliwell that evening, discussing their letter, wondering whether they should take further practical steps. After he had written to his mother, enclosing a cutting of the letter, he left his house.

When he arrived at the Halliwells', trying to restrain his face from breaking into a broad smile, he sensed in his friend something amiss.

"What is it? Is one of the children ill? Margaret?"

"Everyone's well. It's good to see you."

"You seem grave. What's happened?"

"Nothing. Margaret and I were talking. We'd like to meet the famous Marian. We thought perhaps a dinner. Could we arrange it?"

"But something has happened, at your factory?"

"You have not seen the newspaper?"

"Our letter was published."

"But you didn't see the leader?"

They walked to the parlour. Halliwell handed Leggate *The Examiner*, pointed to the first of the paper's editorial articles.

PURE WATER FOR ALL

While public discussion of great municipal works is always worthwhile, we must accord the letter by Dr Leggate and Mr Halliwell, published in this number of *The Examiner*, some consideration before extending it unqualified welcome. It is, of course, gratifying that Dr Leggate, who has chosen to take up practice in our town, believes he has discovered the mode by which cholera is transmitted, and we look forward with interest to his communication to *The London Medical Gazette* so that we may evaluate his findings as they deserve.

Meanwhile, we have to express reservation concerning the conclusion that Dr Leggate and Mr Halliwell have drawn about the adequacy of the new municipal waterworks. Being new to our town, Dr Leggate might not be aware, in his medical reflections on the subject, that the waterworks were built expressly to give our town a constant supply of pure water, irrespective of tidal floods to which this region is subject. Every person in the town will remember that, at considerable expense, the town council retained the offices of Mr Taylor of London, who is known to be the foremost authority in the land on the best methods of suppling wholesome water to metropolitan regions. At the time some doubts were raised about the scheme. Any persons, however, who have taken the trouble to make the short journey to Blackensee Reach, and have taken one of the tours of inspection of the waterworks which have been offered to the general public, can have no further doubts. Ours is the most modern, and the most sufficient, supply of water in England.

Some may feel that we should not have gone outside the town, to Birmingham, for the pumps for this scheme, and some may be aware that Mr Halliwell's workshop in

Bewick Street makes some of the finest pumps in the world, so that it might be argued that his pumps should have been used for the present scheme. It would, of course, be quite improper for any to suggest that not having been awarded the contract for the pumps of the waterworks, Mr Halliwell is now critical of the scheme on that account. He would not be suggesting a new plan unnecessarily because it would require further pumps. No: that is not what we believe. The logic of the suggestion made by Dr Leggate and Mr Halliwell is of a different kind. It rests upon the idea that water from the new works might not be pure. This is patently absurd. The water is taken off from the river when only fresh water is coming down, and after settling and efficient filtration, it is pumped into a noble tower, one hundred and sixty feet above sea level, ready to supply every house in Middlethorpe. Only hold to the light a glass of this water, or taste it, and each one of us may see how mistaken is the idea that the water could in any way be adulterated.

Dr Leggate and Mr Halliwell are to be commended for their concern on our behalf, but scarcely for their understanding of our advances in building new waterworks, and not at all in their attempt to spread unjustifiable alarm; on the contrary, we believe that the hygienic measure of building new waterworks for the public benefit puts Middlethorpe in the forefront of modern towns.

Leggate had read in silence, then remained silent.

Finally he said, "Warrinder—he's taken a dislike to me. Now he misuses his position in this way."

"We miscalculated. We didn't take into account that the mayor, the aldermen, those who subscribed, have faith in the scheme—they're the people Warrinder talks to. They brought in an expert from London, spent a mint of money. The scheme's modern, efficient, it supplies quantities of water—now we find fault with it."

"But we have proof."

"We might as well say that when the cholera reaches Hamburg, a blockade should be mounted at the mouth of the estuary so that all trade with the Continent is stopped."

"Some people will say exactly that. Blockades haven't worked, but we have new evidence."

"You and I know that evidence is important. I'm saying that shipowners would resist a blockade, whether or not there was evidence. Evidence is like putting a farthing on a scale to balance a horseshoe. We didn't estimate the public feeling."

Leggate was cast down by this reverse, but not as much as he might have been. He read again in his *Encyclopaedia Britannica* the life of Leeuwenhoek, who first saw animalcules and planticles beneath the microscope.

It took Leeuwenhoek years before he dared write to the Royal Society, thought Leggate. One must expect opposition, especially from the ignorant and self-interested, and especially if it's anything important. That's why there are special scientific journals and societies, where only truth and facts matter.

Leggate was buoyed by his engagement, by the freer relations with Marian that it brought, by the anticipation of marriage. He enjoyed the idea of moving to a more substantial house. Though he would not have voiced this, he looked keenly towards increased respect in the community, as well as to the practical benefits of a quiet room for a study. He wrote in his diary:

Her family is well known in the town, so we shall be quite set up, established, and when recognition comes, even more so.

He looked on the reverse he had suffered with *The Examiner*'s leader not as a defeat but as a warning. He wrote:

I must be especially exact, utterly compelling. In the letter to *The Examiner* we were too cavalier. Instead of presenting evidence so that it would convince, we merely asserted. The newspaper of a provincial town is not the place for such matters, there is no reason for them to understand. It was good to have this setback before sending the article to *The Medical Gazette*. I shall be on guard, it is that article that will make the difference, and there I shall present such evidence that readers will draw the correct conclusion. But I need more evidence, so the adversaries cannot come back at me. Magendie was always collecting evidence, more and more, he never stopped, the very quantity of it became convincing. What do I have? Two pieces: five houses in a rural parish and a map of a provincial town's water supplies. But what if someone even now has observations like my own, someone with pen poised? Should I not publish what I have?

A week after Warrinder's editorial Leggate found himself no further forward.

Are my thoughts too much on Marian? he thought. Or is it because my practice is growing too fast, and all the time that takes—but it has a good side, because despite Warrinder, who is an ass, his editorial didn't put people off, people have told me that what Halliwell and I wrote is plain good sense, including Anderson who was on the Sanitary Committee; it's decent that he went out of his way to tell me that Warrinder has got above himself.

When I send my results for publication in *The Gazette*, Leggate's thoughts continued, I want them recognized, not mocked. To attack the most modern waterworks in Yorkshire—I should have realized that bigwigs in the town wouldn't like it, and Warrinder does nothing except reflect their views—I should've known better. The medical press and London won't be concerned with local squabbles, for them I must merely be exact.

Nevertheless Leggate continued to ruminate on dismissive editors of contagionist temper and on sarcastic professors of physic who for years had held to the miasmatic view, who would—on the instant that they read anything ill-judged—see the flaw that would negate the whole endeavour.

Why else had Magendie himself not taken up the idea when it stared him in the face? Leggate thought. He must have understood its implications better than anybody.

Leggate remembered the blistering review in *The Lancet* of Magendie's work in physiology. It had accused Magendie of cruelty as a vivisector, said he reached "premature conclusions," it had mentioned "extravagant speculation" and "ill-founded self-confidence." The medical press seldom stinted itself in its criticism. Leggate had read the review before he first went to work with Magendie. If his uncle had not already made the arrangements, so that it was too late to draw back, then on the strength of that article, Leggate would not have gone.

Magendie can withstand attacks, thought Leggate. Established, arrogant, he meets it all with fury. But a newcomer with a new discovery gets one chance—that's how it works. If I publish and it's judged unsound, I might just as well have thrown all my notebooks into the river.

Several times in the course of his engagement Leggate redrew maps of Ringers Beck, of the town's water supplies that showed their correspondence with the incidence of the disease in 1832. At one stage he had half the article—so he thought—then compared it with articles in *The Medical Gazette*. Already this half was too long for one article.

I must be more sparing of words, he thought, confine myself to the important points. But if all the evidence isn't presented, the reader won't be compelled. Should I write to the editor? Ask whether a longer article would be preferred? Perhaps approach a publisher and write a pamphlet. How does one do that? it's expensive no doubt, and money'll be needed for furniture.

Chapter Thirty-seven

Lingering over breakfast, Marian said to her brother, "So you're ready for Canada?"

Marian and William were alone in the room. They fell into the style of talk in which they allowed an affection for each other to come uppermost.

"All fixed up," said William. "I shall leave a few days after your wedding. Two fledglings leave the nest at once. Whatever will Mother and Father do?"

"And have you been finding out about the place?"

"It's quite exciting, a new land, limitless opportunities . . . I suppose it's what I need, instead of kicking my heels round here until Father's gaze happens to fall on some property he can approve."

"You don't sound sure of Canada."

"One can't be sure, but I shall try."

"I suppose it's inevitable, at our age, that we should go different ways," said Marian. It was another strand of apprehension that seemed to wind itself around her as she thought of her approaching marriage.

"I'm not planning to go for ever, I'll stay until the autumn, probably, or perhaps for a year, if Uncle James doesn't get sick of me. But what about you? Mrs Leggate! How will that be? I shall find it hard to think of you as a married woman. Will you like being Mrs Leggate?"

"Why shouldn't I?"

She saw her brother eyeing her. She wondered if he were imagining those new secret relations with a man, in which, as a brother, he would find it hard to contemplate her involvement, or if he were thinking of her public position as the wife of that Dr Leggate who became embroiled in disputes with Warrinder that were published in the newspaper.

"I'll be able to help John in the town," she said. "He doesn't know very many people, that's why he trips over people like Warrinder."

"Has all that upset you?"

Marian looked at her brother, grateful that he recognized the pain Warrinder had caused.

"It's upset me a good deal."

"It's not John who's at fault," said William. "Nobody thinks so ... most
people think Warrinder's too full of himself—too fond of making pronouncements. I must say I much prefer the idea of your doctor as a brother."

Marian looked at him once more, mentally thanking him for his solic-
itude.

At one of Mrs Brooks's Thursday-evening gatherings, Warrinder arrived.

Marian went up to him immediately.

"How could you write that spiteful article?" she said.

"The article to which I think you refer was unsigned, why assume I
wrote it?"

"It's your newspaper, you must have sanctioned it."

"I believe Dr Leggate's and Mr Halliwell's attack on the purity of our
water supplies to be misplaced, it undermines public confidence, and the
newspaper must stand for the truth."

"You didn't reply to the letter I last sent you," said Marian. "I take it that
you don't forgive me."

"I thought I would see you, to speak to you in person," said Warrinder.
"And indeed here we are seeing one another. I do forgive you, completely. I
felt hurt, that's all. It's not unnatural to feel hurt, but it's in the past. The
water business has nothing to do with it."

"I had quarrelled with Warrinder," Marian explained to Caroline, when she
saw her friend the next day. "He said the quarrel had no bearing on his spiteful editorial, but I am sure it had. You can imagine him sounding out the
sanitary committee, and the mayor, to convince himself he was in the right."

Caroline looked grave.

"I didn't mean to quarrel with him," Marian continued. "I hadn't seen him
since my engagement, then I saw him in George Square three weeks ago—I
should have told you—but I said something about the hanging of the man
he had pursued. It touched a nerve in him. He wrote me a strange letter,

reproaching me that when he had proposed, I deceived him by saying that I'd decided completely against marriage."

"You'd said often enough you wouldn't marry; I remember you told him that."

"He says I abandoned a principle . . . he took it as a slight."

"He was content for you not to marry him so long as you didn't marry anyone else, so that he could feel some proprietorship of you."

"Then he misunderstood me," said Marian, knowing her friend said aloud what she herself had felt about her sense of liability towards Warrinder, to which she had so recently been willing to concede once more.

"I should have done more to heal the breach," said Marian. "I felt so furious that I only wrote him a brief note. But instead I should've gone to butter him up, let him feel he still had some special place, assure him of my friendship."

"Perhaps you should do that now."

"After the way he behaved towards John?" she said. "Then he came to the house, to my mother's Thursday-evening gathering—I don't know how he had the gall to do that. I spoke to him very directly about the editorial. If I'd said any more, I should certainly have quarrelled with him irreparably."

When Marian again asked Leggate if she could help with his research, the atmosphere again became strained.

"It upset me, rather," she said to Caroline. "I suppose he thinks it unsuitable for a woman to contribute to science."

"So he won't talk to you about his work."

"He likes talking about general principles of medicine and physiology—the body's a kind of machine, he says. To start with he explained his research, but now he'll only talk about it in vague terms. Instead he asks whether I won't be proud of him.

"I don't even think he's got the paper ready to send off to *The Medical Gazette*," Marian continued. "Though when I ask if I can help, he says he doesn't need help, and that the paper is almost ready."

"Men don't like accepting help," said Caroline, "they think it reflects badly on them. They like to do what they want and pretend they have done it all for you. Peter's always going on about his becoming a Member of Parliament; it would be the most frightful nuisance, but I say I'd be pleased. I suppose I would be pleased for him, but he wants to think that he feels nothing of it except that I'd be delighted."

"You think John feels awkward discussing his research?"

"It's like when you offered to put in a word with Warrinder, on behalf of his letter—it questions his manliness."

"But I'm only trying to help."

"He wants to admire you, and for you to respect his achievements. What if he showed you what he had written, and you found some flaw in it?"

"So I should let the subject alone?"

"At least until you are married—relations are different then."

Relations will be different after marriage, thought Marian.

She wanted to prepare herself, and while her father was away for a day in York, she covertly consulted his *Encyclopaedia Britannica* to see what information she could find. She spent several hours at the task, starting with "Marriage."

> Marriage is a contract entered into by two persons of different sexes . . . is represented . . . as having taken place under the immediate direction of the All-wise Creator . . .

Then in the next column:

> By an act of Parliament of Scotland, 1600 . . .

No good at all, she thought.

She read here, then there, turned pages under several likely headings, was about to give up when she found an acknowledgement of the issue, under "Physiology":

> . . . the seminal fluid . . . the most remarkable circumstance of its composition is the constant presence of an

immense number of microscopic animalcules ... Leeuwenhoek claims the merit of having first discovered
them ... Doubts were at one time entertained ... removed
by Spallanzani.

Extraordinary! Research again, she thought. Like what he sees down the
microscope? So these give rise to conception? And here:

The act of sexual union is prompted by instinctive feelings, experienced in both sexes, and which generally
depend on the condition of the body, and of the geni-
tal organs in particular, which are then in a state of high
excitement ...

And here:

Sect. IV—*The female system*
... receives ... the seminal fluid ... where it can be evolved
and properly nourished ...

"Can't you tell me about it?" she asked Caroline the next day. "The act of
conception. Without circumlocution. I want to know what to expect, so as
to do the right thing."

"It's not something one talks about."

"Don't be unfair—it's something that you know, and you would keep me
ignorant!"

"My mother said it was something that men like, and that it's best to go
along with it, because it keeps them manageable," she said. "But it's not that,
not at all. I think ... I'm not sure." Caroline looked downwards perplexed, a
blush spreading up her cheeks.

"Well?"

"I want it now almost as much as he does, it's ..."

"You forget, I know nothing whatever about it."

So, accompanied by some giggling, and some earnestness, and some incredulity on Marian's part, Caroline put aside her modesty to tell her friend, more or less, what she might expect. And Marian, rather than being shocked, felt excited by this secret prospect. Indeed on the occasions after this talk with Caroline, when she and her betrothed were alone in her music room, she surprised him rather, and increased her own ardour, by lingering over kisses, on occasion by pressing herself rather too frankly, one might think, against him. He responded eagerly, and although the spring weather was not really warm enough, ten days before the ceremony she wore a dress which had a bodice that encouraged her husband-to-be to slip his hand inside it.

The plan was that after the ceremony at St Michael's, and the wedding breakfast in the Assembly Rooms, Dr Leggate and the new Mrs Leggate would entertain Marian's mother and father and her brother, William, together with Leggate's mother and uncle who would come up the day before from London, to dinner at the new house the couple had taken in Petersgate. On the day following, the couple would travel to London with Leggate's mother and uncle. They would stay a day or two there, and then set off on their wedding journey to Germany. A few days after the wedding, William would set off for Liverpool, and for Upper Canada to learn farming with his uncle.

A week before the wedding, a celebratory party was given by Caroline and Peter Struthers. Mr Brooks excused himself, and Mrs Brooks asked if she might instead bring her sister, Miss Mitchell.

A small group stood and conversed. Miss Mitchell was saying, "Marriage seems such an honourable estate that it makes spinsterhood seem quite the reverse."

"You say that to see how many matrons would offer to change places with you," said Mrs Johnson. "There are certainly days when I would consider it."

"Is this the talk for one who is just about to join that estate?" said Mrs Brooks.

Marian was standing in the group, seeing that the occasion invited a raillery at which she was supposed to smile.

She noticed that just three feet behind her, Leggate was conversing with Caroline. She decided to concentrate on their conversation rather than the one that was taking place in the circle in which she stood.

"Your researches," Caroline was saying. "You won't mind being interrupted by the wedding?"

"I've a paper to send off," he said, "but I'm sure it can wait until we return from the wedding journey."

"Of course it's not the most important thing now," said Caroline. "But afterwards, will you write about the drinking water?"

"There are sporules, tiny unseen seeds of the disease," he said.

"Marian's told me your theory, with the seeds of cholera getting into the water. I'm so excited for you to become famous. And you remember you said you would tell me how to avoid the disease."

"Whatever you do, you mustn't drink the town water if the disease returns," he said. "Get your servants to fetch water from one of the springs, Tafts Spring would be all right. And wash your hands very carefully, and get your cook to do the same before preparing food."

"And does this go for all diseases?"

"It may do. For cholera at least, when people become ill with it, millions of the sporules go into the sewers, the sewage goes into the river, then twice a day on each flood tide, it will all be swept upstream, up beyond the water-works."

"So even at low tide there'll be enough sporules in the mud to mingle with the water coming down."

"Exactly!"

Marian heard the eagerness in his voice.

"Your letter to the newspaper before Christmas," Caroline said, "explained the principle very clearly."

"You saw the editorial as well, no doubt. But I did an experiment. I dropped straws that I had coloured with cochineal into the river at one of the sewage outfalls from the beginning of the flood tide. Pieces of straw were still coming down on the ebb during the time they take off water which we drink."

Marian, still three feet behind her betrothed, had become yet more intent on their conversation.

"Better still, listen to this," Leggate was saying. "I analysed some of the town's drinking water for salt. If it's salty, that means the fresh water coming downriver at the bottom of the ebb is still being mixed with salt sea water from the previous flood tide."

"And with it the sewage, you mean."

"Exactly—there's quite a bit of salt in the drinking water, so that proves it."

Marian excused herself from her little group, not too rudely she hoped, and joined her friend and her betrothed.

"Should you like something to eat now, some refreshment?" she asked.

Reflecting on what she had heard of Leggate's and Caroline's conversation as she lay in bed that night, Marian tried to resist the dull pain she felt. She knew the signs—she had experienced them clearly enough in Leipzig, when Andrei had announced his engagement to the fair-haired, silly Grete.

Is friendship a fortress to which I retreated, she thought, so that I should never suffer this annihilating jealousy? Now I feel it towards him, with whom I'm to spend my whole life and towards Caroline—the two people to whom I am closest. Why does he talk to her of his work, and not to me? And then she says how excited she is for him to become famous!

Marian exhaled audibly. I must think I own him, she thought. I'm as bad as Warrinder, worse—John was only talking to Caroline about his letter, and the river. But I can't help this feeling. After all those declarations about being candid with each other, in a few weeks, what do I do? I store up secrets: Warrinder's proposal, the reason for his spiteful response to John's letter, now this appalling jealousy.

Chapter Thirty-eight

Five days before Marian's marriage her father fell ill, with his usual complaint of chest and throat. He said he would be well enough to take part, but on the day before the ceremony he was not, and his brother from Harrogate agreed to take his place. On the morning of the ceremony, Mr Brooks seemed worse. Mrs Brooks and Marian went to see him, Marian in her wedding dress so that he could see her looking lovely, and feel included. Newcombe advised that it would be best to go ahead, and hoped Mr Brooks would feel sufficiently well later in the day for Mrs Brooks to visit once more, and to bring the newly married couple with her.

In St Michael's Church Marian looked up into the vaulted roof, felt herself tiny beneath it, as she walked slowly forward. My father not here, she thought, I shouldn't be doing it without him, how can I be doing this at all with him so ill? . . . should I be doing it at all? now I give myself over to this other man—from one man to another, like a kind of chattel. Is it what I believe? But . . . there's no drawing back, I believe I do love him. And in the service, when I say I shall obey, I shall make it clear to God that in our case this is no part of it.

The newly married couple and Mrs Brooks visited Mr Brooks again after the service. He had not improved. Marian pressed his hand. He was only able to say faintly: "So it's all done? You should get back to your guests. Come and see me tomorrow."

That evening, when the guests had gone, it was a sombre party at Dr and Mrs Leggate's dinner table. While toasting the newly married couple no one could avoid thinking of Mr Brooks.

"He may be starting to feel better by now," said Leggate. "These fevers, after four or five days, they often reach a peak, and recovery begins. Why don't I go and find Newcombe, and go and see Mr Brooks, and bring the news."

"I wish you would," said Mrs Brooks. "Dr Newcombe went to see him as the guests were leaving, but all we got was the message, by way of Lucy, that there wasn't much change—perhaps I should come with you."

"You stay, if there's any change, I'll be back."

Leggate's plan may not have been a good one—he knew Newcombe would be doing everything possible—but in the state of anxiety that the little party was in, it seemed sensible.

It was an hour before Leggate returned. He looked grave.

"I can't pretend to be hopeful," he said. "I've talked with Newcombe, who is with Mr Brooks now. He may pull through, but he has pneumonia on both sides, and a very serious fever, and—Mrs Brooks, William, Marian—it may be best for you to go to Sutcliffe Street now. There's a chance . . . a chance he won't last the night."

There was a shocked silence—what everyone feared was suddenly present in the room.

"Go straight away," said Leggate's mother. "Your uncle and I will go back to your house, John, as we arranged, and we'll see you in the morning."

She went to Mrs Brooks, kissed her kindly on the cheek, and said: "You are in my prayers." To William: "And you too." To Marian: "And you, my dear."

At the rectory it needed no medical knowledge to see that Mr Brooks was dying. Mrs Brooks and her two children sat together with him for an hour, then decided that each should spend a few minutes alone with him, first William, then Marian, then Mrs Brooks.

When Marian was at her father's bedside she said, "Can you hear me Father?" He turned his head, a fraction of a movement. "I'm sorry, Father, that I have been disobedient, and wilful . . . can you forgive me? and bless our marriage, I know it's not ideal as far as you were concerned, but if you could . . ."

Mr Brooks, too weak or his throat too affected to speak, made a sound which Marian at that moment took for a "yes."

"I love you, Father. I pray for you, and may the angels bear you upwards." She kissed her father's burning forehead. In tears she ran from the room.

Chapter Thirty-nine

Next day was a day of arrangements. Leggate discussed with his mother whether she would stay for the funeral—she would not think of doing anything else, she said.

Leggate sent telegrams to cancel hotels and the steamship booking. He went with William to see what was necessary for the funeral—William cancelled his journey to Canada.

Leggate suggested Marian try to sleep in the afternoon—she was unable to do so. But during that time he went to sit for an hour with Mrs Brooks.

It was not late on the evening of that solemn day that he made his way, with his new wife, to bed.

"Shall I blow out the candle?" he said.

Neither of them had slept since the night before the ceremony.

"Not yet, leave it, I don't want to be in the dark. And with all this, on our second day of marriage we've not even properly married, in the flesh."

"No," said Leggate.

"I want to, I have wanted to," she said. "You know that."

"We don't need to until you are ready," he said. "We can wait. It's all been too shocking."

"It's not the shock . . . it's that I know you want to, but I won't be able to, I feel too wicked, it will intrude, and be the worst possible basis, for our first time."

"Wicked?"

"Wicked, wicked," said Marian. "I didn't love my father enough, I was disobedient."

Her head rested on his shoulder as she lay by his side. He felt the material of his nightshirt damp with her tears. "And I can't stop thinking about it long enough for . . . for anything else . . ."

"It's all right," he said.

They lay in silence.

"Take off your shirt," she said. "I want to rest my cheek on your chest, to feel you there. Will you be able to love me even though I'm wicked?"

"I do love you."

But this brought more tears, dripping directly onto his skin.

"I will come to you, I will—I have to find a way to stop these thoughts, these thoughts," she said. "If I could only get them to stop. Here, I want to feel you properly, and for you to feel me. To show I'm not separating myself from you."

With this Marian slipped her arms, first one and then the other, out of the puff sleeves of her nightdress, sitting up to give her husband a moving sight in the candlelight.

"Remember when you put your hand here," she said, taking his hand and putting it in the same place again. "I felt overcome with excitement. Just let me rest on you, like this, for a while." She nuzzled herself against him. "If I can get these thoughts to stop."

I suppose they must have stopped, thought Leggate some time later, as he realized she was asleep. What a muddle.

Chapter Forty

In the Leggates' house in Petersgate the shutter for the centre panel of the window of their new bedroom was cut to leave a nine-inch gap at the top. It admitted the first light of morning, enough to greet any who woke early but not enough to rouse the sleeping.

On the second morning of their marriage, the faint rays of an overcast dawn slanted through this gap, making dimly visible to Marian the face of her husband, nestled into a pillow.

To have him here, she thought. She saw his eyes open. She smiled, then remembered... My father.

She said: "Last night. I must have fallen asleep. I'm sorry."

She allowed herself to be drawn to him. She felt him close. She was able to banish thoughts of her father. She pressed herself towards him. Though his body felt larger, stronger than hers, it was reassuring. I could nestle here, she thought, like an infant.

She felt his hand on her neck. She felt him drawing her towards him. She inched herself closer, as if she might merge herself into him. But there's nothing of the child in this, she thought, it's this which is the very secret rite of adulthood. I must be ready, not reluctant, not distracted.

"Let me get rid of this," she said.

Her nightgown was still bunched around her waist. She took it off quickly, and nuzzled again against him. She felt his presence, strange but warming.

She heard his voice: "Shall we do this now? Would you be able to be with me?"

"I want to. I'm thinking of you very much."

"You're not frightened?"

"Caroline gave me very complete instructions."

Aware that she still was mentally holding off a spectre that, if she were not vigilant, would overwhelm her once more, she suppressed too an impulsive thought of her jealous reaction to Caroline a week ago, and with it the thought that Caroline was the kind of woman men wish to marry.

Am I really the kind of woman men wish for in this moment, she thought, or am I here on a train of chance, and really he would desire someone of that other kind, fair-haired, soft, feminine?

"Do you really and truly want to?" she asked. "Are you sure?"

Knowing that he could not fully understand the meaning of her question, she needed still to hear a certain answer.

"Very much."

She felt his arms surround her, felt his fingers tease the hair at the nape of her neck. It was the answer she needed, that would make this possible.

"You're not to worry the slightest bit that it might be painful for me," she said. "I know it can be uncomfortable the first time for a woman, but I shan't mind. Promise you won't think of it."

She kissed him now. Remembering Caroline's instructions, after a little while she reached secretively beneath the covers to discover what her friend had told her of, previously sensed only indistinctly through thick layers of cloth.

Then she lay to receive him. She felt just a little discomfort at first, not really a pain at all. She placed her hands on his shoulders—what an extraordinary thing this is, a kind of interpenetration of selves. Then, feeling a strong movement within him, and a little sigh, she felt a moment afterwards his face come to rest against hers. All at once she was overwhelmed. This thing, they had done it. It had moved him in this strange way. Now he lay upon her, like a baby, on her breast. She placed her hand on his head, felt a sense of protectiveness, that she would do anything for him, represent him to anyone, then her eyes started once more into tears.

Marian scarcely ate between the wedding and the funeral. At night she hardly slept. Worst of all on two occasions she had shaking fits, in which she felt such extreme apprehension that she could not imagine any feeling in the world could be worse.

On the first of these, her husband was at Sutcliffe Street with William. But on the second, she sent for the maid, who went then to fetch him.

"John," she said. "How can I stop this, it must be the entry to Hades. It shows I've been wicked."

He held her hand. "I'll give you a draught," he said. "To soothe, and help you sleep, but it'll take some time to have its effect. I'll go and get it."

"Don't leave me now, don't leave," she implored. "I can't stop trembling, it's completely outside my will . . . and shaking. What is it? Something unspeakable is going to happen, I know it, and my heart, am I going to die?"

"A kind of fit," he said, holding one of her hands in one of his, pressing his other hand to her forehead. "It can happen when one's had a shock."

"I'm going to lie on my front," she said, turning over. "Just lie on top of me, cover me over with yourself, completely."

In ten minutes she became calm. "Am I going mad?" she said, as her husband rolled to her side. "Tell me truthfully."

"Not madness," he said.

"Then what? A fit, apoplexy—don't keep it from me."

"A fit of dread," he said. "If anything . . . anything with a Greek name, a hysterical fit."

"And shall I choke?" she asked. "Is that what happens next? It's caused by the womb—is it because we did that thing, when we should not have, because of my father . . . And you'll send me away and have me locked up?"

"It's not really anything to do with the womb," he said. "That's just a name, and a superstition, and of course I won't send you away. What you should do is to rest, or take walks, or play your piano—or if you like, I'll read to you."

"Is that what you recommend to all your patients?" she said. "Reading?"

She did not want to be read to; but she wanted her husband close by— he would sit and hold her hand, or stroke her brow, and when she got up she said the last thing she wanted was to go out. All she would do was practise Chopin's Marche Funèbre, which she insisted she would play at her father's funeral.

Marian was surprised to see Warrinder at the funeral; he offered his condolences.

"Mr Warrinder! Whatever are you doing here?"

"Mrs Leggate." He emphasized her new name. "To pay my respects, of course. Or you might say I came to attend an important event for the newspaper."

"You never liked my father."

Warrinder made an offhand gesture. "So the shock of your marriage was too much for him," he said.

Marian glared at him. "Are these your respects?" she said, and turned her back.

Marian was sitting silently three weeks later. This she did for much of the day, unable to read because she could not attend to the words sufficiently. She rose only to pace the room. In the evenings she would fall asleep at last, because she was exhausted, but was wide awake at half past two, unable to sleep again.

Her Aunt Mitchell made her fourth visit to her niece. "Now I'm the one sitting at the sickbed," she said.

"It's no good reading to me, I'm sorry, I can't concentrate, I'm sorry, Aunt."

"Don't keep apologizing. You've had a bad shock."

"You think me weak, marrying, being looked after by a man?"

"I think you're ill."

"What will become of me?" Marian's face, a month earlier so fresh, was now thin, and apt to crumple.

Miss Mitchell placed her hand on her niece's arm. "This too will pass," she said. "At the darkest moments, this is what you must remember: it will pass."

Marian gave her a tear-sodden look, not disbelieving so much as uncomprehending.

Marian experienced only one more fit of trembling after the funeral.

"So I am not going mad, and my womb is sound," she said to her husband, after it seemed there were to be no more such fits.

"There was never a question of you going mad," he said. "And I believe I was wrong about hysteria, independently of whether that state has anything to do with the womb. I think you were suffering the first stages of melancholy—not unusual in a bereavement. I've talked to your mother. I suggested I take you away."

"You're sending me away!" Fear spread across her face.

"I said 'take you away,' both of us—the journey we couldn't take before. A journey is the best prescription in these states. We'll go to Leipzig, as we planned."

Marian would receive Caroline, who had come frequently to visit.

"He thinks it will take my mind off things," said Marian after her husband had prescribed the journey, "but I can't imagine it will. I've fallen into a dark pit—all the ideas I had before were quite wrong, about friendship, so I did not equip myself. You were sensible, you looked forward to being a wife, and are happy, a mother, a bulwark for your husband. All I am is an invalid."

"There's nothing wrong with friendship. You and John will be friends."

"You don't realize, I fixed on friendship because I've not the constitution for the other, I'm not capable of love."

"You talk such stuff and nonsense."

"I'm not capable—when Andrei preferred that wretched German girl I felt jealous, jealous. I can't support these passions, they're too much for me, they make me someone I don't want to be. And a week before our wedding, when John was talking to you at your party, talking about his science in the way he refuses to talk to me, I felt jealous of you too. Love sets off passions that I can't restrain, now they've unhinged me. I married against my father's will. My mother mourns in the ordinary way, but I'm deranged."

Marian, encased within her own thoughts, had not spoken to anyone with frankness since her father had died. Now came this sudden outburst.

"You take it too much to heart," Caroline said. "And you have it exactly the wrong way round; John was speaking to me as he would to a man, to Halliwell. If anything it's most unflattering to me as a woman. He wasn't flirting. Can't you see? He doesn't want to discuss his research with you until it's published and acclaimed. He wants his wife to see it only when it's perfect."

"You think so?" Marian looked at her plaintively. "You must explain these things, because I can't think straight any more. My thoughts are in the most terrible turmoil."

"John's right, a journey will do you good, just the two of you."

Leggate talked to Caroline next day.

"I'm glad you're taking her away," Caroline said. "Everything has closed in on her. She spoke to me yesterday in such a way ... it frightened me."

"Saying what?"

"Saying that her thoughts and passions are all in a whirl—said she was even jealous of me."

"Jealous?"

"Of when you explained to me about your river experiment, before the wedding—she thought we were flirting."

A cloud crossed Leggate's face. "Delusion," he said. "Even before the wedding."

He remained pensive for the space of a minute, with Caroline waiting apprehensively for his next pronouncement.

"She's very withdrawn with me," said Leggate. He was not used to confiding such matters, but felt on this occasion that it was not inappropriate. "She just apologizes all the time, says 'did you know what a mistake you were making in choosing me for a wife?' and suchlike. She says she feels cut off, as if enclosed in a glass jar, able to see people outside, moving about and getting on with their lives, but not able to touch them."

"You must talk to her," said Caroline, "or rather, let her talk to you, to say aloud all these things that are tormenting her, otherwise they just go round and round in her mind. She tormented herself with this feeling of jealousy for me for nearly a month."

Leggate and Marian, with Mrs Brooks and William, in due course gathered to hear Mr Brooks's will. William had spoken to the solicitor, and arranged to make the reading of the will a short affair. The solicitor read the relevant passages, some minor bequests to servants, and a sum for the refurbishment of the church organ, and then generous settlements on his wife, and his daughter. Marian had not imagined her father had so much money. Then, as was proper—and Mr Brooks strove always to be proper—they heard that the residuary legatee, heir to his father's estate, was William.

"Mr William and I have a good number of matters to discuss," said the solicitor. "Financial matters... Mr Brooks's affairs... there's a deal of property, I don't think this is the place to discuss it all, no doubt, er..." The solicitor looked at William. "No doubt you'll want to discuss it all with your mother when it's all sorted out."

"Is there anything we should know, that we should be concerned about?" said Mrs Brooks.

"Nothing, nothing," said the solicitor. "I wanted to let you know you were well provided for, it's just that your late husband owned quite a bit... quite a bit, and it will take some time to get everything in order, but nothing to be concerned about, quite the reverse, the family is very well off."

With his new responsibilities and his financial independence, William's character seemed to change. He would stay in Middlethorpe now, he said. He would see that his mother was properly taken care of, the first thing would be to arrange a new house for her. And if, later on, a farm came up, not too far away, with no hurry of course, he would consider it.

Mrs Brooks, though subdued, bore bravely her part as widow. She would look for a house in a month or two, she said, and no, she did not want to come and live with Marian, that would not be fair, but she would certainly live nearby.

Chapter Forty-one

Marian said, "We could go to Scotland, they're not having a revolution there, are they?"

She was responding to her husband's reiterated suggestion that they make a journey. But the idea of going on the Continent only made her apprehensive, particularly since that spring of 1848 had been a time of revolutions in Europe, as if a train with open firebox had rushed from country to country scattering sparks of rebellion on a tinder of discontent.

"I've always thought," she said, seeing if she could persuade him, "that I'd like to visit the place near Glasgow where Robert Owen set up his ideal community . . . and it wouldn't be as far, or cost as much, as going to the Continent."

She added this in the dejected voice of apology that Leggate had come to dread.

"But you're well off now. We can afford it."

"I don't deserve it."

In the end he prevailed. They would take the steamer to Hamburg, then to Leipzig, and to Dresden—these places seemed safe enough, and Leipzig's familiarity to Marian helped to make the idea seem possible.

The crossing was calm. Marian and her husband started to fall into the way of travellers, thinking of where they would arrive, remembering what they had seen, reflecting, talking sporadically.

"Do you think me wicked to marry?" Marian asked, as she stared across the sandy waves of the North Sea. "No, don't answer, you'll only say something to humour me."

"Not wicked," he said. "It's generally considered one's duty to marry."

"I spent half my life defying my father. Marrying was too much; I think it killed him."

"In your state thoughts are exaggerated. Your father gave his blessing, you said he did. He and your mother both did."

"I hope he did. It was my last request, when I saw him on my own, when he was dying. I asked directly ... he made a noise, but I don't know if he gave his blessing, or refused it. I think now that he refused."

"Is this what has been preoccupying you?"

"And other things." Marian started once more to weep.

"It was an accident. He'd been ill in the same way before. It's to do with tiny sporules, probably all fevers are caused in a similar way—not anything you did."

"Warrinder said it was. He said the shock of my marrying killed him."

"Warrinder! ... Why Warrinder?"

Marian lapsed into silence. At length she said: "You don't think my wickedness killed him?"

"I think microscopic sporules grew in his lungs."

Marian sat in an armchair, looking through a second-floor window of a solid hotel in Hamburg, that solid prosperous city. "I wonder where Herr Mendelssohn's family house was," she said to her husband. "He was born in Hamburg."

"We could try to find out," said Leggate.

That perked him up, Marian thought, probably has it marked down as "patient showed sign of reviving curiosity in the outside world." Would I want to see Mendelssohn's birthplace? she wondered. Hard to imagine it would be of much interest.

"You talked of Warrinder during the crossing," Leggate said. "As if he had something to do with bringing you down."

She saw him look startled by her glance.

"I'm sorry," he said. "I shouldn't have mentioned it."

"No," she said, "you should. It concerns you, concerns us."

She stared through the window a long time. "You know he proposed to me?" she said. "A long time ago, three or four years ago."

"No ... you didn't tell me."

"I meant to tell you, then everything got tangled."

Should I tell him? she asked herself. What good would it do? I behaved dishonestly, if other women can be straightforward in refusing proposals, why not I? I led Warrinder on, then I wasn't honest enough to admit I had been in the wrong. I am justly punished.

"You're angry with me," she said to her husband.

"I'm . . . I don't know what to think."

"I was wrong, you see, I am a deceitful person."

"Deceit . . ."

"I should have spoken." She was silent for another long period, then said, "When Caroline became engaged, Warrinder and I were thrown together, he's her cousin, you know, and he became attracted to me, I think, well I know he did, because he came to the house, and asked my mother if he could see me, and wanted to propose, and I made him a promise."

"You what?"

"I said I couldn't marry him. I'd given him the article I had written, on friendship, to read—the one I gave you—he read it and he proposed, so I told him I could never marry, that the only thing dear to me was friendship, just as I'd said in the article, but that I could promise to be his friend."

"So there is something between you?" said her husband.

"No . . . yes. No longer. I did it to fend him off . . . not usual, I know, for a man and a woman to be friends, but I believed it then. So you see, I was wicked, I didn't keep my part of it—because now I've married you."

"And that's why he's so hostile to me?"

"It's hard to avoid thinking so."

She looked towards him, searching his face, seeking some reassurance that he might understand her predicament without her having to retrace every sordid step.

"After you and I became engaged, I met him," she said. "Saw him in the street, I mean, and he wrote a hateful letter, accusing me, and I apologized and said . . . that's why I couldn't tell you, I said I would continue to be his friend."

She saw her husband rigid, as if paralysed. "A pact," he said.

"No."

He did not reply.

"See—you are angry! Rightfully so. See what damage I have done?"

"I'm trying to understand. You don't love him?"

"I don't love him, I never have and never could, that's the very reason all this happened."

Marian saw her phrases had a healing effect. The furrows on her husband's brow smoothed, the muscles round his mouth loosened. He lifted his clenched hands from the wooden arms of the chair on which he sat, and laid them quietly on his lap.

"You've been keeping this to yourself?" he asked, not unkindly.

"Caroline knows a part, not all."

Leggate slipped away from the hotel. While his wife slept, he would get some fresh air. He walked alongside canals and dismal warehouses, across prosperous squares, along heavily built streets.

So Warrinder could contain himself so long as she never married, he thought. And when we became engaged, he was furious, and at the funeral, at the funeral can you believe? he cruelly accuses her of killing her father, by marrying. All the talk of wickedness over the last month is of wickedly deceiving him, of wickedly giving way to the temptation of marriage, of wickedly defying her father.

Caroline's warning came to him again, as it had repeatedly in the last days, that Marian had too many matters eddying in her mind.

Like an abscess, he thought, but one that doesn't drain, and after dressing it, next day the purulent matter's there, as before.

Leggate, struggling still, as he had been during the past weeks, to keep his own self from the despair of disappointment, directed his mind to untangling what had occurred. How often I remember thinking she must be promised, he thought. So why didn't I think that she'd received other offers? Of course she must have.

Though his thoughts were reasonable in themselves, it was as if they staggered, slipping, and with the sound of tumbling gravel, on the edge of an aching fissure that threatened to pull him in. Warrinder, he said to himself. Of all people, Warrinder!

I know his kind, he said to himself, his mind continuing to work over the issue, as if by comprehending with precision the type of person this was, the pain that had been caused would evaporate. Command, that's what such people want… to have sovereignty. When their subjects are subservient, they can be benevolent, even generous, but once stray outside their will, then only cruelty. Magendie had something of that… Why was she attracted to such a man? Making up to him. Leggate could not expel the image of the brute Warrinder, on a bed, his gross naked body pressing upon Marian.

But at last, coming in a circle once more, back past pointed-roofed warehouses along a canal, the image faded.

Now, he said to himself. Now what? All those women I've listened to, every one with something pent up. Now my own wife. And what of me? How am I affected by knowing this? Am I now married to an invalid? And have I now learned to hate the bully Warrinder?

Marian was awake when her husband returned to the hotel. She was sitting once more in the armchair, gazing through the window, across the tree-lined square.

"You hate me now," she said.

"I love you. But we must start on a different basis. Without all these unspoken things between us."

"As real friends," she said.

The word "friend" was not the best one to have chosen at that moment, and as soon as it was out, she longed to reclaim it. But it was too late. It was off in the air between them. It was she who had dispatched it. Marian looked at her husband as he moved to stand behind her chair. She looked up into his face, and as she did so, he leaned down to kiss her on the brow.

How can he bear to kiss me? she thought.

"I don't want to be wicked," she said. "Don't let me be wicked."

She took his hand from her shoulder and laid it on her breast. For the hundredth time, tears started in her eyes.

Leggate was pleased his wife showed occasional sparks of interest in the world. They went for a walk together in the streets of Hamburg.

"Lessing lived in Hamburg," said Marian. "One of the town's great men."

A little later, passing a theatre: "We could go to the theatre," she said. "Or a concert. I think I could manage that, shall we?"

"A concert would be better for me. I couldn't understand a play in German, and I've dropped the theatre, for years now."

He shocked himself, hearing his own words, having not until this moment thought that he should tell her of Evette. I've said we must start from a new beginning, he thought.

"What are you thinking?" she said.

"I dropped the theatre because of a painful memory," he said. "When I was in Paris. Here, let's stop and have a cup of coffee."

In a mahogany and brass restaurant, in a shadowy corner near where waiters pushed through the door to and from the kitchen, Leggate reached out to take his wife's hand on the tabletop. "There is a past for me as well that I should have told you. I didn't want to say anything to upset you."

"A love affair?"

Leggate stared at the table, and nodded his head very slightly. He felt his wife's hand become tense.

"Years ago, I was only twenty when I met her."

"But you didn't marry."

"She was an actress, she said she would never marry." Leggate noticed his wife look startled.

"But you would have?" she said.

"It was before I knew you."

"But you did that thing together?"

Leggate nodded again, almost imperceptibly.

"Over a long time?"

"For three years . . . nearly four."

"She was fair-haired?" asked Marian.

"She had black hair."

"So you can be deceitful and immoral."

"I should have told you."

"Does that make us equal?"

"Should you like to be equal?"

"I don't know," she said. "I don't know. You must give me time. To get used to it."

Marian was very silent that evening, and the next day.

"Perhaps it's for the best," she said, at last, in the hotel restaurant, at dinner. "It makes us equal." Although she still left her plate almost untouched, her mind would sometimes now work according to its more natural rhythms and modes.

"How could I expect you not to have fallen in love before?" she said. "You're not incapable it seems . . . I could scarcely have been the first. Were there others?"

"No others. I'd put all that aside, until I met you. Hoping to meet . . . someone like you."

"Hm," said Marian. "I don't want to be thinking of her, when we . . . and I don't know whether it's best for you to tell me everything, or nothing at all . . . But I don't suppose you want to be thinking of Warrinder, either."

"Ask me just so much as you want," he said, "and I'll be candid. And about Warrinder, has he been the only one?"

"There was one before," she said, curious as to how this had gone quite out of mind. "Not a proposal, the reverse, he didn't propose to me but to someone else."

Leggate remained silent, waiting.

"Andrei . . . I loved him," she said, simply.

After a pause. "In Leipzig. We played in the orchestra, and we would talk . . ."

"Is that why we're going to Leipzig?"

"No!" said she, loudly, surprised. "It's because I wanted to show you the place, and for you to hear the Gewandhaus Orchestra, and so that you could see the conservatorium . . . Andrei wasn't interested in me as a wife. He proposed to his landlady's daughter, the doll-faced Grete."

Two married people retired to their room after dinner that evening. With curtains drawn, candles casting shadows, the couple climbed up, one each side, into the heavy oak-framed bed and burrowed under the eiderdown.

Marian said, "So many ghosts in the bed."

Leggate said, "How to perform an exorcism?"

"It was a stupid kind of ignorance, the worst kind, thinking we were the only people in the world."

"Can you forgive me?" he said.

"I like to hear you say that, so it's not only I who've been wrong."

"You've not been wrong . . . we're together."

"Exorcism," she said. "We must do that thing."

They embraced, then taking off nightgown and nightshirt, they began to kiss more strongly.

"Will you be able to think just of me?" she asked.

"Will you of me?"

"You are not fair—they've never been here, Andrei or Warrinder. You forget, for me this has only ever been with you."

"I think of you continually. And of how glad I am to have found you."

"Not recently. You haven't looked glad, only distraught and dutiful."

"But underneath."

"How I hope so."

They lay on their sides, face to face, tentative fingertips gently made progress across a brow, along a shoulder, down a collar-bone. And when a certain point had been reached, Leggate prompted Marian to roll onto her back.

"That Evette," she said. "What would she do? I want you to show me."

"So you're allowed to think of her?"

"I don't want you hankering . . . is there anything?"

"We could . . . like this. I'll lie, and you . . ." Leggate guided his wife to sit upon his thighs. The eiderdown had slipped from her back, and she sat before him. "And then, you see . . ."

"Caroline didn't tell me . . ." but she leaned over him, and kissed his lips, and he held her hips, to show her how to move.

She sat up. "Was she a harlot, this Evette?"

"She was an actress, making her way through a difficult world, no different from you or me."

"And you will just be able to think of me?" she said, leaning over him, touching his lips once more with hers, and touching his chest.

"With no difficulty."

And in a little while there occurred in him that inner transmutation for which he sighed, and which she had come to recognize. She smiled, and bent down to kiss him once more.

She snuggled beside him. "Don't sleep," she said. "Was that . . . ?"

"Mmm."

Each saw the other's face on the pillow, eyes looking at the other, each seeing in those eyes a companion on a journey, but neither one knowing route or destination.

"There's something else," he said.

"There is?"

"Lie still."

And kissing her gently, flicking his tongue at her ear lobe, he reached down and with his fingers invited her legs a little wider, and moved gently, caressing, detecting a certain readiness, a willingness indeed, expressed in a movement towards him, and kissing her more deeply now, at last there occurred that involuntary trembling, prolonged, as if on a brink, and still, and trembling still and then, a small sound.

She held him tightly. She was weeping. A strand of hair, wet from tears, stuck to her cheek. He moved the damp strand to her brow, stroked her forehead. "Are you all right?"

"I'm all right. These are good tears." Then after an interval, the tears ceased and she said, "How little I've known. Is that as it is with you?"

"Caroline didn't say?"

"She didn't say."

Each one lay quietly, thinking, thinking . . .

"I keep wondering whether I'm capable of it," Marian said after a long interval. "Constitutionally, I mean, of being able to tolerate love, that's why I thought that friendship was best, not passions, because without them one doesn't have to bear such torments."

She looked at her husband's face. He looked towards her. They seemed to each other within reach now.

She touched a finger to the little groove that ran from his nose to his upper lip. "But you, you're not to make me jealous, ever."

"I'm very loyal."

"It's why these feelings are kept to marriage," she said, "otherwise they just turn evil, because they're too strong . . . And that's the knowledge that Adam and Eve must have found. They discovered that, from this act. Too strong . . . they cut people off, even from God. And that's what made them feel ashamed."

"You're getting better," he said. "Back in books and thoughts, again."

"Do you mind that? You want me to be someone else?"

"I don't want you to be anyone else."

Chapter Forty-two

Leggate took rooms for a week in Leipzig. At the Conservatory, Marian arranged that she could practise there in the mornings, and during this time Leggate would explore the town.

She's still very labile, he thought. For an hour or two, there's her spark again, then down, down, down. Was all her insistence then, about friendship, a kind of fear? At least it was enough to fend off that Warrinder.

On their second afternoon in Leipzig, Leggate found he could contain a certain thought no longer.

"When Warrinder wrote that article in the newspaper in response to Halliwell's and my letter," he said to her, "was he doing it from spite, because you'd become engaged?—You said he wrote you a letter."

Marian cast her husband a mournful look. "I had goaded him," she said. "It was my fault. I saw him one day, on an errand, just after the hanging of that man Biggs. He said he was squeamish about seeing Biggs's death, and I cruelly said that it was he who'd put the noose round his neck . . . that's when he wrote me the letter, and I wrote him back. I said I'd been beastly to say what I did, and said I was abjectly sorry, and said again that I'd be his friend . . .

"Not long after that," she said, "you remember, you sent your letter to the newspaper, and I said I would talk to Warrinder to see it was properly displayed, but you wouldn't let me. To do what I offered, I'd have had to humble myself to him again . . . I would have done it, but that's why I couldn't tell you, when I should have, and why getting married seemed . . . a betrayal."

"A betrayal."

"Of what I believed, of renewing a vow of friendship to him, binding me further, as I shouldn't have done when I was an engaged woman, because of my weakness . . ."

"So I was a weakness?"

"Everything got into a terrible knot."

Leggate had asked permission to visit the botanical garden of the Leipzig Medical School. He took his wife to walk in it.

As they left the garden, he said to her, "Your father? How was he involved in all this? Why do you keep saying you feel wicked in relation to him? It's to a family's honour for a daughter to be married. I'm a gentleman."

"I don't know," she said. "He disapproved of me. Something went wrong... it was he who taught me the piano, and to read..."

"And you loved him."

"For a time in my childhood, until I was about thirteen, he meant a great deal to me..." Her eyes filled with tears. She looked appealingly at him. "I can't talk in the street, everyone will see me."

They walked back silently to their lodgings.

"What my aunt told me," she said when they were indoors, "is that when I was thirteen, I don't know exactly when, he started not to like me, and to say I was defiant, and unnatural."

"You wouldn't bend to his will, and he started to bully you."

Marian looked alarmed at such a word. "One must respect one's father... Honour thy father and thy mother... I tried, but my defiance of him was too strong for me, and marrying was part of it."

What a very little thing that makes me, thought Leggate. He did not know how to respond.

"And so, when he died, and when Warrinder said that thing at the funeral, that my marrying had killed him, it was as if the whole force of my wickedness..."

Leggate looked at his wife, tears running down her crumpled face, abject once more. Her tears dissolved his indignation.

"You must have felt utterly forsaken," he said, "when you were thirteen, and your father seemed not to love you any more."

"Aehh..." Marian made an inarticulate sound. "Unloved..." she said, starting a fresh torrent of tears.

Leggate rose, gave her his handkerchief, for hers was now soaked. He knelt beside her chair, and took her head upon his breast.

The couple attended concerts and, as they had arranged, they saw the conservatorium. Two days later they took the train to Dresden, where Marian had reserved a two-hour lesson with Clara Schumann.

When she arrived at Clara Schumann's music room, Marian looked at this woman, barely five years older than herself, half a lifetime advanced in experience; wife and helpmeet to her famous husband for eight years, already the mother of four young children and of one who had died in infancy, and, among all this, one of Europe's most celebrated pianists—the pianist Marian most admired. With her there were matters to settle that could not be decided with anyone else.

"I'd like you to play for me," said Marian. "I want to hear you, not in a concert, but just here, the two of us."

"But you have reserved a lesson."

"And then I would like you to hear me," said Marian.

"Very well," said Frau Schumann. "What shall I play?"

"First, anything you have written, a sonata... anything. And then I would like it if you could play some Chopin, anything of his that you play, it would be best really if it's something that I also play, for I play a good deal of Chopin... I want to be able to hear you in something of your own, and then in something I know well, so that I can compare with myself."

For almost an hour Frau Schumann played, and Marian listened.

"It's inspiring," she said, at last. "Though dispiriting... you're so perfect."

Frau Schumann looked at Marian, and did not reply.

"I dread to ask," said Marian. "But could you hear me, now? There is only one thing I have been playing since my father died, Chopin's Marche Funèbre."

"Your father died, my sympathies."

"Just two months ago."

Marian played the march, sombre, stirring...

"It's good," said Frau Schumann. "But I could advise a little..."

For fifteen minutes she did advise, and demonstrate. "You see now?" she said.

"I do see... thank you, but what I wanted to ask is this. When you hear me, do you think that, if I practised more than I have been practising recently, with my father's death... that I could reach the concert stage?"

Clara Schumann looked taken aback.

"I want you to be utterly honest," said Marian. "I need to make a decision about my life, though I see already from your look what the answer must be."

She continued to regard Frau Schumann steadily.

Marian spent some hours in contemplation after her lesson with Clara Schumann. There's a woman, she thought, whose talents are extraordinary and who, among the minutiae of the busiest domestic life imaginable, has achieved a partnership with her husband, even despite his so-obvious neediness towards her and the fact that he would be visibly jealous whenever she attracts more attention than he. But their life's shared around music. Because his hand's injured he hasn't been able to play in public, so instead he has composed. She puts herself behind his composition, which is his strength, helps in writing arrangements, conducts rehearsals—most of all she plays his works in public. She composes as well, but in a subsidiary way; she's not well known for her composition. No, the pattern's a shared domestic life, and two modalities: he writes, she performs.

Marian said to her husband, "I've made a decision. In some unspoken way, before I married, I'd thought . . . you see I didn't imagine I would ever marry . . . but I'd give concerts, perhaps, not like Frau Schumann, of course, who is fêted everywhere, but in small places around Yorkshire . . . to give me something to aim for."

Leggate waited, fearful of what she might say.

"But I've decided . . . I decided instead, I won't give up music, just play for myself, for you, and for friends, but I've decided. I want to put myself behind you and your researches, to assist you in whatever way I can, and to be a wife."

IV.
Doloroso

The Coming of the Cholera

The streets were dull and quiet, with the exception of the now constant tolling of the passing bell, warning people how busy death was amongst them, and the melancholy procession of bearers carrying away the bodies to the great pits prepared for them, unaccompanied by any mourning friends whatever. All business was suspended; the vessels in the docks lay at their moorings with their great tall white masts standing up grim and silent like spectres from the spirit land. The people went about with fear stamped on their pale faces, as they hurried out of each other's way, and not a cheerful note of any kind made the universal gloom brighter; but one prolonged wail of sorrow went from every heart—even from the little band of helpers who had their number diminished in this great struggle between life and death, without any being willing to supply the vacant places.

> *Mrs Curd*, Ellen Carrington:
> A Tale of Hull During the Cholera in 1849.

Chapter Forty-three

Returned to Middlethorpe and to her new house, Marian began her new life.

"I was ill," she said to Caroline. "Confidence, hope, everything, all gone. I didn't know of such states, or myself capable of them."

"You're still very thin. Do you feel recovered?"

"Not really recovered. I still wake and dread the day, but by the afternoon I'm more determined. I've not suffered much before in my life, then Warrinder saying I had killed my father, and the fact that I had gone back on telling him I wouldn't ever marry, and feeling I had been wicked. It all somehow demolished me."

"Not marriage itself?"

"Not marriage. You've always said it was good that I should marry, and John . . . I don't think he could have been any better."

"Don't think?"

"It's not his fault. He expected a wife. Instead he got another patient . . . and we are close sometimes, but not how we were when we used to practise songs together, when we were engaged. He sees me as a new responsibility."

"It was all an idea," said Caroline. "You must see—the way you would talk about friendship. I can't imagine why you let yourself get so immersed in it. You care for John, why can't you just believe in that?"

"Friendship was something I cherished. I betrayed it, propelled by 'chance and sensation,' and now I'm much more dependent than any woman who ever lived. I'm shown up as no different from everyone else."

"You're no different from me, then," said Caroline, in some exasperation.

"Don't be cross. I did believe in something different and more equal. I'm not saying I should be different from you—but can't you see? It didn't seem impossible at the time. I was thinking all in the wrong direction, so what's resulted is that I made myself completely, utterly, unprepared."

"I want to shake you," said Caroline. "All this morbid introspection."

"It does go round in the mind." Marian looked anxiously at her friend, to see if she had exhausted her patience.

Caroline asked tentatively, "And that intimate part, which in marriage is special?"

"I was looking forward to it, with what you'd told me, but... I have nothing to compare it with..."

"It can take some time."

Marian gave her friend a plaintive look. "You think so? I've let him down there too."

When Marian spent time with her mother, they conversed as easily as before her marriage, though of course now everything was utterly changed.

"I'm not going to let it get me down, mustn't feel sorry for oneself," said Mrs Brooks.

Is Mother dropping hints? thought Marian. I suppose she's right, I'm wrapped up in myself, sorry for myself.

"I wanted to say that I feel for you, Mother, about Father."

She took her mother's hand.

"I was too distraught before I went away, and I don't think I told you properly, my sympathy... and I miss him too," she said.

"Thank you, my dear," she said. "You and he didn't get on in recent years, but you did when you were younger."

They held hands a little while longer, until Marian kissed her mother on the cheek.

"You're looking for a house, Mamma?" she said.

"William's been very good," said Mrs Brooks. "In a crisis, he did better than expected."

And I've done worse, thought Marian.

"You could stay with us," Marian said. "I've told you that."

"I'm not going to impose," said Mrs Brooks. "The new rector won't be here for a while. You needn't worry about me, it's you we need to worry about."

Marian shrugged.

"I've never seen you look so thin and haggard," said her mother. "Perhaps I should come and stay, and see you eat properly."

"I am eating, Mamma."

"I don't believe you—you were always uncompromising, and now you've taken this hard. You must think of your husband, you know. Life must go on."

"Don't reprimand me, Mamma."

Her mother sighed.

"Did Father accept my marriage? . . . truly I mean, not just acquiesce?"

"Of course he did."

"When he was dying, when each of us went to see him alone, I asked him to forgive me for having been rebellious, and asked whether he would give me his blessing. He couldn't really talk—just made a noise, and I don't know whether it was yes or no."

"How can you think such a thing? When we both saw him before the ceremony, when he wasn't so weak, you know he gave his blessing—quite explicitly, you know he did. I heard him myself."

"I didn't think of that . . . it was that last meeting which stayed in my mind . . ."

"You should remember it—that's when he was really with us, that's what you should remember. If you weren't a grown woman, I'd spank you."

Seeing her daughter's distress, she rose from her chair, and put her hand on Marian's forehead. "You're still not at all well," she said.

And then, as if coming at last to a purpose that she had been postponing: "We need to have a talk, though. You and I and William, about your father's affairs."

With sudden apprehension she looked at her mother. "You mean William's found something out, and we're destitute?"

"No," said her mother. "Your father was wealthy, as the solicitor told us. William wants to discuss with us what he should do."

When he met with his mother, and with Marian and Leggate, William seemed to have become master of his new situation. As heir to his father's estate, he had no obligation to discuss matters with his womenfolk, but he had decided, nonetheless, that it was the decent thing to do.

"I've had quite a few meetings with the solicitor and with Father's agent," said William. "Here's the thing. Father inherited a good deal from Grandfather, and it turns out they owned a sizable piece of land in the town, houses, and two large buildings. Father kept it all quiet, and only the solicitor knew, and everything was done through an agent, you know, Mr Theobald. But anyway, here's the thing . . ." William seemed unsure of how to proceed.

"Here's the thing. One of the buildings we own, well I suppose I own, is number 12 George Square, the offices of *The Examiner*, which the newspaper has a lease on, which is coming up for renewal, and I mean . . . after Warrinder was cruel to John, I mean, by being beastly about the suggestion that he and Mr Halliwell made about the water supply, I wondered if we should sever all connection."

Marian was shocked—that not just lives and sentiments but financial affairs in the town should be so intertwined.

"You mean to sell the building?" she said. "Do you need the money, to buy a farm or an estate?"

"I've got my eye on an estate, the other side of Hadby, a very good house, and a home farm, and not too far from the town here. But I don't need to sell anything, I was asking you in case you felt we simply shouldn't have any dealings with the Warrinder family, throw them out, and let the building to someone else."

"Don't do anything on my account," said Marian. "On our account." She looked enquiringly at Leggate, who shrugged his shoulders to indicate that he thought he should have no say in the affairs of his wife's family.

"If Father could keep it all quiet, then the arrangement can continue, as far as I'm concerned," said Marian. "And that's all right with you, John?"

"Of course."

Marian gathered from his glance at her that for him there was nothing at stake.

"I shall keep it quiet, then, just as Father did," said William. "Nothing much need change, and as far as old Mr Warrinder and Edward Warrinder are concerned, they just deal with the agent, though I think I'm going to appoint someone new there. Since I've looked into things, and made some enquiries, I've come to the conclusion that Mr Theobald wasn't much good as an agent, or not as good as he should have been, so I'm thinking of getting in someone else, but everything will be done still through the solicitors, and through the new chap."

"It's entirely your decision," said Marian. "John and I are grateful to be asked. It's kind of you, but please don't let sentiment on our behalf weigh at all, especially since it's all going to be done through an agent—I can't see it changes anything."

"We'll keep on as we are then," said William. "Though I'm going to up the terms for the new lease, and make the lease an annual one. The building's ideal for them, the newspaper's got a very good position there, with their printing presses behind. It's the natural place for them, in the centre of the town."

He looked at his mother. "Is that satisfactory to you, Mother?"

"It's in your hands," she said.

"And if you wanted me to see if it would be possible to buy that house you're in so you won't have to rent," William said to Marian, "I could look into that for you. You could use part of your settlement on it, if you wanted, or you and I could go shares. Whatever you liked."

"Perhaps," said Marian. "It's not urgent. But thank you. John and I will talk about it."

She looked wonderingly at her brother, the man of affairs.

"Thank you for asking me," she said again. "I'm grateful. It makes me feel cared for."

Chapter Forty-four

Leggate wandered through the rooms of the house in Petersgate, thought of settling into some mode of seeing patients and spending evenings conversing with his wife to keep her spirits up.

I must find a way forward, he thought. How did it all so quickly vanish? My discovery, my new wife—and what have we now? Abundant scratches in a notebook . . . and responsibilities.

He went to his new study, and sat before a handsome desk. She is recovering though, he thought, the journey was good for her, but I didn't dream her constitution was so fragile, she seemed so full of life and strength, that I never dreamed, then suddenly snatched away—a glimpse from time to time, as if to remind me, and then a wraith. Even if she recovers, it will always be there, a susceptibility, needing utmost care, with a constitution so frail. Many women have it, it's my job to look after her.

I suppose with Evette, he thought, she was older, with a far different temper, more like an adventurer. I was so young. She took care of everything.

Leggate made a motion with his hands as if shaping a vase, then repeated this containing gesture, and stared at the imaginary object he had created.

I imagined all that would be regained, he thought, but that's not the way, now it's my turn to do the taking care, the containing, or perhaps it's more like passing something on.

He linked his fingers together, stretched his arms above his head. Passing something on, he thought, perhaps that's the way to conceive of it.

He considered how he should get started again on the impossible business of sending off an article. Or should I start again? He sighed. I know my conclusions are right, but neither I nor the world is affected, not the slightest bit, by what I've found. In any case I must go and see Newcombe . . .

"Everything was turned upside down by her father's death," he said to Newcombe, after thanking him and his assistant for looking after his patients.

"The weather's not been bad here, not too many coughs, and your patients not too importunate—they'll be looking forward to seeing their favourite doctor again." Newcombe smiled.

"I don't know whether to get started on researches again. It was a habit, and now it's broken."

"Of course you should—you owe it to the world. If you don't, I shall come burgling and steal your papers, and publish them under my own name."

Leggate looked at Newcombe wanly.

"Tell you what," said Newcombe. "You kept saying you'd join the Medical Society, but were never keen enough to do it. Let me introduce you. We meet on alternate Wednesdays. Anyone introduced speaks after the ordinary presentation—don't make it too long. I'll see if I can fix it up for the Wednesday after next, and you can talk about the five farmhouses on the stream, and the correspondence with the water supplies in the town before the new waterworks was built. It'll start off your thinking again, you'll do the paper for *The Medical Gazette* in no time."

Leggate worked on his paper to give to the Middlethorpe Medical and Scientific Society. He compressed the essentials of his discovery so that it could be read in twenty minutes: "Researches on the propagation of cholera in 1832 in Middlethorpe and its environs." Following the experience with *The Middlethorpe Examiner*, he decided to present only the evidence that deaths were closely associated with supplies of water, and the likelihood of their contamination from sewage, cesspools, and drains, but not to discuss his theory of sporules. His paper was received with attention. The chairman thanked him, asked for questions or comments.

Dr Gadd rose. He looked like an inquisitive heron with a habit of peering across the top of his spectacles in a way that increased the effect of his considerable height.

"Gentlemen, we have heard a more than usually interesting rendition. I offer three remarks.

"First, we are told by Dr Leggate that the unique cause of cholera is water that has become contaminated. He mentioned the idea in the newspaper some months ago, and now he has passed round a map showing deaths from the cholera in 1832—quite fascinating. I commend Dr Leggate for his researches, which he has conducted with diligence. Perhaps the quality of water fifteen years ago did contribute to the disease. But without disrespect to you, sir," he glanced towards Leggate across the top of his spectacles, "I wonder whether the paper we have heard should not be regarded as preliminary. I wonder whether we should not, for the moment at least, resist the speculation that contaminated water is the sole means of transmission of cholera. It is known that people contract the disease from an atmosphere of muck and filth of every kind, from the squalid conditions in which they live, as well as from immorality and vice. All these are well known. Cholera is such an appalling condition that we in our profession must strive to combat it. Even suppose that contaminated water contributes, for us to imagine it to be the sole cause would be to avert our eyes from the manifold evils of filth and immorality. It would be dereliction of our duty.

"Second, Dr Leggate proposes that the physiological substratum of the disease is alimentary—the intestine. Again, Dr Leggate may be urging us in a direction that is too selective. In 1832 I well remember one man who had no diarrhoea or vomiting, who died within a few hours of falling ill. The medical literature attests to other such cases.

"Third, our profession requires us to minister to the sick. Should not any aetiological theory prove its worth in therapeutics? Dr Leggate's researches have not as yet, as I understand them, reached the question of how to treat the afflicted."

Leggate replied. "Thank you, sir. I am grateful for your remarks. No doubt filth and squalor do contribute, I fully believe they do. My research is directed to finding what part squalor might play, or rather, what . . . what the active agent is in the squalor. This summer the disease is sweeping through Russia, approaching as we speak. I do not deny that filth affects the disease . . . But the question is this: 'What is it about filth that does the evil work?' I suggest it's the filth in drinking water that has the effect, and if so, this will help us know how best to act when the disease returns."

Four more men spoke. They took their cue from Gadd. None discussed anything Leggate had said. One agreed with the first speaker, that immorality and vice in the lower classes certainly were the primary causes of cholera. Next an elderly physician rose to tell of his experience of cases in the epidemic of

1832, describing first these, then those, then other symptoms. Two final speakers rose to give their suggestions for the effective treatment of the disease in question, and of disease in general.

Leggate made replies which became shorter and shorter, trying to keep exasperation from his voice.

The society was not of venerable foundation, but it had robust traditions. At the conclusion the chairman rose, and laid out the rules.

"As you know, gentlemen, it is our custom, after a paper has been read and discussed, to ask a guest to withdraw. The members of the society then vote on whether an abstract of the guest's paper be entered in our proceedings, and whether you, sir, are invited to join the society." He looked towards Leggate. "The secretary will communicate with you tomorrow. Once again we thank our guest this evening."

A desultory fluttering of applause, then with the eyes of the chairman upon him indicating that nothing further would occur until he retired, Leggate left the chamber.

It was after eleven o'clock when Leggate reached home. Marian sat in their drawing room reading a novel. She was dressed in her nightgown, a shawl on her shoulders. She looked up. "I waited up for you."

"Thank you, my dear."

"Was your research well received?"

Leggate could not bear to tell the truth—should humiliation in public be brought indoors to be multiplied by her knowledge of it?

"It went well enough." He crossed the room, kissed her lightly on the forehead, then slumped into a well-upholstered armchair.

"But what . . . there's something you are not saying. Was someone argumentative?"

"Argumentative?" Leggate held himself back from roaring at her, as he would like to have roared at Gadd. "Yes, argumentative! Benighted, small-minded, parochial—the whole town, the whole rustic gang of them—yokels!"

Leggate caught his wife's eye. "I didn't mean that, not everyone in the town, not Halliwell and Newcombe, not you, but . . ."

"Who was it, who was unkind?" said Marian—she went to sit on the floor, taking her husband's knee under her arm.

He did not answer.

"Wasn't Newcombe there? Was it anyone in particular?"

"A man called Gadd."

"Oh, Gadd—he's very full of himself, doesn't like to be contradicted. He's the moving spirit in the Medical Society, he was behind the Sanitary Report."

"No doubt a friend of Warrinder," said Leggate.

Marian looked hurt. "They know one another," she said. "It's a small town—and they can be small-minded." She pressed Leggate's thigh to her breast. "Don't let them wound you."

Marian realized, at that moment, that he would need her. It was as if, in a river, the water that had been flowing in one direction was now, with the change of tide, perceptibly flowing in the other. This was not how she had imagined it. She saw with certainty that there would be times when he would need her!

With Father's death, she thought, we set off in utterly the wrong direction. I won't be an invalid—I'm not cut out for a pianist on the concert stage, but I can certainly be a proper wife, we'll see if I can't.

A defiance which previously, as it seemed to her, had no proper object, rose within her. I'll see about that Gadd and all those others, she thought. I'll make sure my husband's work is recognized.

Leggate, next day, received an early call.

"I'm most dreadfully sorry," said Newcombe. "I miscalculated. I can't begin to say how much I..."

"The Medical Society doesn't accept me," said Leggate.

"That's not it exactly: I expect you'll get the secretary's letter in the second post. I did what I could, Anderson spoke up for you. The meeting was stormy, I hadn't anticipated..."

"It was Gadd."

"'Suggestive but not yet conclusive,' those were his words."

"No one could be more conclusive. I have evidence."

"Gadd's not a bad fellow. He does take up these lordly positions. He's the power behind the Medical Society, was one of the founders. Responsible for that building. People defer to him."

Leggate was silent.

"Gadd said the research was highly commendable but as yet preliminary, that he would welcome a further application in due course—then I think that was accepted as the wording."

"They don't know how far I have come."

"It's the Sanitary Committee that made the report on the condition of the town. They can't accept there's anything they didn't cover."

"I was polite."

"I thought they'd be delighted to welcome a clever researcher. They will welcome you, I'm sure. I'll make another appointment, and you can speak about therapeutic and preventative issues in cholera. Just say something bland, anything . . ."

"After humiliation I am to crawl back?"

"Don't think like that. Gadd is protecting his position. Thing is . . . I'm sorry I didn't think of it, but I should have gone to see Gadd beforehand, to get his blessing. Then there wouldn't have been a fuss. I'm sorry. Just think of this as a delay. There are several on your side."

"Not enough."

"Gadd and the Sanitary Committee believe they have a firm grasp of the problem, that disease occurs from the whole condition of filth and stink, everything, all together. Your idea makes much of what they've done irrelevant."

"It is irrelevant. Middlethorpe Committee of Hypocritical Bigots."

"Don't despair. I'll discuss your admission with Anderson and some of the others. I'm sure your next talk will be met warmly. I blame myself . . ."

Silence.

"You should write your article for *The Medical Gazette*. You must have it by now. You covered the main points in your talk. It would make a perfect article."

Leggate did not reply.

"London, that's where your discovery belongs, not the Medical Society of a tinpot provincial town," continued Newcombe.

Leggate at last was touched. He knew Newcombe was fond of Middle-thorpe, proud of being a Yorkshireman.

"I shall do the article," Leggate replied.

"Give it to me, when you have a draft. I'll make suggestions, do whatever I can. Would you like me to write to the editor, to Wakley? I don't know him but I've had a couple of things in *The Lancet*."

Lancet! thought Leggate. The blood-letter's instrument. Why can't they even give a serious name to a journal?

Out loud, he said, "I'll try *The Gazette* first. My maps would make Wakley think I'm in the miasma camp."

"I am confoundedly sorry. They behaved like boors."

Marian, at breakfast, asked: "Who was that calling so early?"

"It was Newcombe."

"He brought news from the Medical Society?"

"He did." Leggate wanted only for a large fissure in the earth to open, for himself to disappear within it.

"They're nincompoops," she said.

They ate in silence, until the new girl brought in the post. The letter from the secretary of the Medical Society was there. Leggate had recovered himself to some extent. He opened the letter. It said what Newcombe had supposed.

Marian considered. She decided on plans, both for the immediate future and over the longer term.

In marriage, two people become linked together like piston and flywheel. The piston pushes, the flywheel turns. With its circular momentum the wheel gives energy back to the piston, easing it past those regions where it lacks its own impetus, until the steam again enters the cylinder, the piston again pushes strongly on the wheel.

At the start he was the piston, she the flywheel. The train of the Leggates' marriage had set off, the piston pushed, the flywheel turned, but they were off in quite the wrong direction. Marian found herself in just that place she deplored, the weak, dependent woman, sustained by a dependable man. They had to stop, reverse, go back to their station, before they could begin again.

Chapter Forty-five

Contemplating her new state as married woman, Marian was galvanized by Gadd's treatment of her husband, following Warrinder's malicious squib. She did not ruminate, for it was not vengeance that activated her. But with her strong core of loyalty, she identified herself with her husband's struggles. She concluded that the part of wife that she had taken up was not ignoble. She would bear children, make a family—and she would certainly see her husband was taken seriously.

Enough debility, she said to herself, awaking at eight o'clock in the morning after a long sleep, a few days after the Medical Society's disgraceful letter. She was surprised to feel no longer, as she had every morning for the last ten weeks, the despairing dread of another day. Her new determination had made her debility dissolve.

I refuse to be an invalid, she thought. I don't know what came over me. I shall certainly not go on like that.

At Caroline's, Marian was pleased to see her friend.

"You're more cheerful," Caroline said.

"I feel more determined, or sometimes at least. Everything was upside down. You were right—marriage is a perfectly satisfactory state."

"I wouldn't say perfectly satisfactory."

"Now you take the other side?"

"Keeping house is not enough to sustain one, you need something else, you especially."

"I shall help John get his research published. There's something that prevents him. And in other ways too . . . be damned to the sick-list."

Caroline looked surprised at this military expression, but Marian, granddaughter to a colonel, was not going to apologize.

"You gave me advice about the bed part . . ." Marian continued.

"It didn't hinder you?"

"Quite the reverse."

Marian began to tell Caroline in some substantial detail about her excursions into marital relations. "And do you know," she said . . .

"While you were at your patients," said Marian to her husband when he returned home, "I played the piano." Until that day she had not played since her lesson with Clara Schumann in Leipzig.

"You mustn't give it up," said Leggate. "And Clara Schumann may not be right. Apart from Chopin himself, you have the finest touch of anyone I have heard."

"I'm not giving it up, and you flatter me, with 'touch'—it's not touch, it's technique that does it, and technique is practising five hours a day, unremittingly from the age of six."

"You could be a concert performer, if that is what you wished. I believe in you." He rose from the table, and kissed her on the forehead.

"Would you like me to read to you?" he asked. "What about the new book by Currer Bell, about a mad wife in the attic?"

"How could you! That was nearly me. Should you like me to be shut away?"

"You were never mad, just in low spirits."

"All the same."

Marian thought: he must think I'm recovered, or he'd never tease like that, it would be too cruel.

"Not reading then . . . a walk?" he said.

"If you hadn't been so heartless as to tease I would have suggested we go to bed."

"It's early."

"All the better."

"I don't really have a weak constitution," she said, when they were in bed, unclothed, she lying on his chest. "I had too many wrong ideas. I so want us to be all right. Together, I mean." She rose on one elbow, looked her husband in the eye, and jostled him a little with her breasts.

"We will," he said.

"I want us to do everything, more than you and that hussy could ever think of."

She felt him turning her over onto her back, and his mouth pressing on hers. "You're not to think of her," he said. "Only of us."

"And neither are you."

They kissed a good deal more. He stroked her stomach, then the nape of her neck, as she lay once more on his chest.

"I read in one of your medical books that at that moment, you know, of the crisis, that the male seed passes into the female."

"You shouldn't be reading those."

"Why ever not? I read it in the encyclopaedia before we were married, and I confirmed it again this afternoon. That is what one is supposed to do when one doesn't understand something. First you read about it, then you try it out, then you read about it again."

"What an idea."

"You can observe the male seeds under the microscope, little animalcules... seen by Leeuwenhoek, though people didn't believe him."

Leggate looked shocked, as if he thought this quite unsuitable matter for the bedroom.

"Could we see them?" she asked.

"How do you mean?"

"See your seed under the microscope—we would have to catch them. I can't know what happens inside me."

"From being melancholy, you become fantastic."

"Why fantastic? You say the body's a machine; now when we talk of its most important powers, you look flabbergasted. I'd like to see them—that's all. You've got a microscope, there must be a way to catch them."

"We could I suppose..."

"Why not?"

"This is about love," he said.

"Yes it is," she said. "About not keeping me in ignorance."

"We couldn't do it now," he said. "There's not enough light. You need good light for the microscope, lamps and candles aren't enough."

"Another time, then, in the daylight. I shall kidnap you from your study tomorrow afternoon, take you away from your precious notes, and you can show me how they can be collected."

"For now I don't want to think about such machinery," he said.

"So orthodox."

"I want our souls to touch."

For Marian, who really was almost recovered, whose excursions into physiology were not just a rebound from dejection, the idea of souls touching was enough for her to put aside, for an evening at least, her researches into microscopic cells and to concentrate on the still-novel investigation of more substantial bodies.

At breakfast some few days after this, Marian glanced towards her husband as he waved a letter he had just received. "We're invited by Lord Andervilliers, to a house party in a month's time, and he asks if you'll play at a musical evening—and the following evening there's to be a ball."

"Should we go?" asked Marian.

"How can we not? You'll enjoy it, perhaps a step in your career."

"I don't have a career," she said. "Except here."

"Perhaps this is how you start one."

Marian gave him a sceptical glance. How lightly he thinks of it, she sighed inwardly.

That week Marian put out flowers, and started again playing her piano for three-hour stretches, felt that she became her own self.

"We can read together," she said after dinner. "I would like to read that Currer Bell, and I'll play the piano for you, I'm still very rusty, you mustn't mind."

"I won't mind."

"And I'll help with your writing—I'm good at it. You don't need to impress me, I want to help make your discovery known."

"It's very technical," he said. "Medical terms."

"And I'm very clever, and know Greek and Latin better than you—poo to medical terms. Can't you see? We've had a reverse, and must get going again, and I can help—I'd like to . . . make a contribution. And you're to tell me where your researches have got to."

Unaccountably, there was a catch in Marian's voice. Leggate looked at her with a frightened expression, as if he would have to start up his apprehension about her again. "I will," he said, "tomorrow. I'll tell you everything."

Next morning, after breakfast, he showed her his maps, the farmhouses on the stream, the traced overlays. He explained the theory of the sporules, told her of the experiments with the river, and with analysing the salt content of the town water. "What do you think?" he said.

"You've solved a problem that people haven't been able to solve in two thousand years," she said. She felt excited by her husband's achievement, felt moved by his devotion to this cause.

"You think it's watertight? No flaws?"

"Your manuscripts," she said. "Just give me the paper you read to the Medical Society and as much as you've done of your paper for *The Medical Gazette.*"

Leggate was reluctant. He moved slowly to find the papers, at last found them, placed them before her.

"You can read them," he said. "I've got patients now, and I've something to find out for my map. I'll see you later on."

"You need not look so apprehensive," Marian said. "I'm going to help, not criticize. We can do it together."

That afternoon Marian was excited for her husband to return. "Here," she said, "I've rewritten some of the paragraphs of your paper—when you talk to me about your discovery, you make it very clear, but the paper's rather ponderous in places, so I've changed some of it, and made some suggestions . . . but I need to discuss some bits with you."

Marian saw her husband's face had fallen. "Thank you," he said, gathering up the papers. "Can we do it tomorrow? I feel tired from my patients, it's too much to have to work on this today."

He took the sheaf of papers, and went to his study.

"I'm sorry," he said later that evening. "I can't do it. All I did today was worry, knowing you were reading my paper, before it's ready. And I was right—you say it's no good."

"It is good," said Marian, "very good, just the forms of expression."

"I know it," said Leggate. "I've read your changes. I'm trying to make what I write better, but you don't realize, for everything one says, one must be careful to say precisely enough but not too much, and not to make claims that one can't substantiate. One must qualify everything so that a reader doesn't get the wrong idea and then attack one ... qualifications are the centre of scientific writing. And for the very same reason there's the danger of being dull, but it's not like trumpeting in the newspaper. If one makes wild exaggerations, as you have, one is utterly ridiculed. This bit here," he waved a paper of her handwriting. "'Sporules that reproduce themselves explain every essential of the spread of cholera'—you just can't say that."

"It's what you believe."

"But you can't say it." He looked agitated, as if he were considering whether shouting would help his argument. "What if there's some essential they do not explain? Don't you see?" He tried to modulate the edge in his voice. "I've not even seen the sporules under the microscope."

They were silent.

"I'm sorry," she said.

"I must do it myself," he said. "There's no other way."

From that moment, a barrier was erected—not between them in their ordinary lives, for after some days of awkwardness following the quarrel over his article, with Marian's new resolution, they reached with each other a relation that was close to what each had imagined before they married.

There was no barrier as she hovered throughout the rooms of the house, seeing if he needed tea when he was in his study; no barrier when she would leave the door of her music room open and for two or three hours a day send through the house the lovely sounds of her piano, no barrier when she asked at dinner time about his patients, no barrier when they would read novels together in the evenings, no barrier in the growing delight they took in their marital bed.

Perhaps "barrier" is the wrong word. Marian saw it was more that a compartment had been constructed, a container, and in it Leggate put his science, keeping it from her until—she supposed—at last after much taking it out to adjust it, and secretly to tinker with it, and then put it back again, he would be able to take it out finally, proudly, in front of her, and offer it to her, fully perfected.

But it's all wrong, she thought, he should talk to me about it, we could think about it together—I've a good mind, could help him see weaknesses so that he could correct them, or... all this stuff about qualifications in scientific writing, he's got it upside down, he thinks that by diluting and qualifying he won't be challenged, but that's not it, he needs to have thought through all the possible challenges. Surely he can only do that by explaining it to someone else, to me, so I can think of objections, then he could get evidence to counter them. But it's no good entwining himself in qualifications; he should write clearly. Is it that he suspects there is some weakness, and doesn't want me, or anyone, to see it?

Marian stared at her hands, perplexed. I'll start reading some of his books, and think about the problems on my own, she thought. There'll come a point when he needs to talk about it, then I'll be ready.

One evening as they finished dinner, and Marian had asked what her husband would like to do, he said, "Can we go over to the Halliwells' this evening? I want to see Halliwell, and it would be good for us to go out."

At the Halliwells', Leggate said to his friends, "Did you see in *The Times* today? There's cholera in Saxony. You have an atlas?"

Marian shuddered. "You know we were there a few weeks ago," she said.

The atlas was brought. Leggate made for his friend a calculation he had made many times for himself. "About six hundred miles to Middlethorpe. In round figures, say that cholera travels at five miles a day. Six hundred divided by five, about a hundred and twenty days until it could be here, four months. Could be less."

"So soon," his friend replied.

"We should warn people," said Leggate. "There'll be panic, calls for quarantine. Some will argue against quarantine because of damage to trade. All the while people could be safe if they stopped drinking contaminated water."

"Should we approach the Sanitary Committee?" said Halliwell. "Four months is time to sink a well at Tafts Spring, install pumps, lay a pipe. It could be done in half the time."

"I've spoken to many people who took water from Tafts Spring in 1832. I believe we can say it's safe," said Leggate.

"Why not go to the Sanitary Committee?" asked Marian.

"It's no good. Gadd is the Committee. You know he thinks me unsound."

Leggate and Halliwell decided to meet next day with Newcombe to plan what to do.

Chapter Forty-six

Next day, Leggate met with Halliwell and Newcombe. They agreed they should think widely.

Newcombe went with Anderson to see Gadd, who was polite. He agreed that there was risk of the return of cholera; he had himself renewed the meetings of the Sanitary Committee when the disease had broken out in Norway.

"Closer to Middlethorpe than Saxony," he pointed out. "The Committee's already met with the town council. There's not enough progress being made with clearing nuisances. That's the thing to concentrate on now."

Finally he said: "A pipe can't be justified, not at the moment. Many more urgent priorities than a pipe from Tafts Spring, even if dirty water does in some undefined way contribute to the disease. If there's money to spare, we should spend it on clearing nuisances."

Newcombe also went with Leggate to see the mayor.

"I know next to nothing about science," the mayor said. "The Sanitary Committee has it all sewn up. But we're on top of it, public subscription for clearing up nuisances and whatnot. We've got a gang started already, they'll be racing along with it. And the new waterworks—cost a fortune. Best in the land."

It was decided, when he met with Newcombe and Halliwell, that Leggate would write letters: one to *The Times*, one to *The Middlethorpe Examiner*, one to Mantell, author of *Thoughts on Animalcules*. These could be done in a day. After that, his main task—all agreed—must be to finish his article for *The Medical Gazette*.

For the letter to *The Times*, Leggate gave the essence of his discovery only. He started thus:

Cholera is on the march once more. In 1845 it was in Kabul, in 1846 it was in Persia, this July it ravaged St Petersburg, now it is desolating Germany, soon it will be upon our doorstep.

He laboured long over these opening sentences: "on the march once more" had the right sound. With martial overtone, he thought, gives a sense of danger. It will catch the eye.

Leggate received a note saying that although the editor valued his letter, he was sorry that there was insufficient space to print it.

Leggate's letter to *The Examiner* referred to his earlier letter in that newspaper, described where people's drinking water came from in 1832, named supplies, streets, specific pumps, standpipes. Leggate explained how those who drew water from some sources had died while others who drew water from sources such as Tafts Spring, or even from the well in the workhouse, had survived. He argued again for a subsidiary pipe from Tafts Spring to the waterworks, in case the disease now approaching so closely should affect the river.

As from *The Times*, Leggate received a note from *The Examiner* thanking him for his letter, and saying they regretted they could not print it.

This second note was even more infuriating than the one from *The Times*. Pure prejudice, he thought. Who do they think they are? I am not established. That's the reason. The town newspaper and the Sanitary Committee all in league—"very good report," the mayor says. "Town can be proud of it," he says. Very proud they'll be when the cemeteries have to close because they're overflowing.

Leggate paced. Letters by Gadd get published in *The Examiner*, he thought.

The bitter recognition that truth cannot demand a hearing came late to Leggate. His work was painstaking, every fact recorded, every inference warranted, every statement duly qualified.

What I've found is not just opinion, he thought. Opinion's the principal content of every newspaper, every medical bag—anyone can have an opinion. I can offer one: "Standing on the head revives a failing circulation." There! Stands to reason. The blood has an easier time running back towards the heart. If I wrote that in a letter to the newspaper, it would be published in an instant.

He continued to pace. His memory of the conversation with the mayor kept returning. "I know nothing of science!" he mimicked. Here's something on which he could exert himself, and what does he do? He refuses. But doctors should know about science. Gadd and the Sanitary Committee are no better than quacks. It's to them that the preposterous Warrinder listens.

Nothing pleased Leggate less than the idea that his intricate chain of discovery, with each link forged, welded, tested, was considered crankish. He thought of writing to Magendie, to enlist his aid. But as he knew well, Magendie held to his own opinion.

And Mantell, the ally who would be pleased to hear of corroboration of his research on swarming sporules with its link to infectious diseases, how did he respond? He had his secretary write a polite note:

> Dr Mantell instructs me to thank you for your communication, and for your generous words on the subject of his book *Thoughts on Animalcules* for which he is profoundly grateful. He begs to inform you that he awaits with keen interest the publication of the researches to which you refer, in *The London Medical Gazette.*

By the end of August, although no article had been sent off, the preparations that Leggate, Halliwell, and Newcombe had made were ready. Halliwell had had six water carts built, huge barrels mounted on two-wheeled carts, each to be pulled by a sturdy horse. A new well was sunk at Tafts Spring, and three new hand pumps installed there. Since the newspaper would not print Leggate's letter, Halliwell had paid for three thousand copies of a handbill to be printed.

PRECAUTIONS AGAINST THE CHOLERA
When the cholera visits Middlethorpe again, as is likely, we will all be at risk, men, women, and children, but it is possible to guard against contracting the disease because, by tracing the patterns of cholera when it visited the town in 1832, Dr Leggate has found that its principal cause was in drinking water that became contaminated, by way of drains and sewage, with the evacuations of those who suffered from the disease. The symptom

of cholera is a plentiful watery diarrhoea. This diarrhoea contains the poisons of the disease; it enters the sewage, and the sewage is transported to the river. The disease spreads, therefore, because its poisons mix with the very same water that we drink. Contamination can also occur if a person touches clothing soiled by the evacuations of someone with the disease, and then eats or prepares food.

You can protect yourselves as follows: if the disease visits the town, it has been shown by experiments that the River Keel is likely to become contaminated so, since this is now virtually the town's sole water supply, you should take drinking water from elsewhere. We will draw water from Tafts Spring which supplied safe water during the last epidemic, and which has no drains or sewage emptying near it. We will make this spring water available from water carts, which will travel through the town. You may bring jugs and buckets to these carts to be filled, or alternatively, pumps have been installed at Tafts Spring, and you can obtain water from there.

Observing such precautions will largely decrease the risk of contracting the disease, so that God willing, we may survive the coming epidemic.

Dr. John Leggate
Dr. Henry Newcombe
Thos. Halliwell, Esq.

"Timing will be important," said Newcombe. "Too long before the disease arrives, and people will forget."

"Sometimes there are sporadic cases before the whole thing propagates," said Leggate. "We would anticipate the worst if we act when the first case appears in Middlethorpe."

"Why not draw a circle round the town, of fifty miles radius?" said Halliwell. "If any case should occur within it, we distribute our handbills."

"A man with the disease could come on the train from London in a day," said Leggate.

They were pensive. "If the disease reaches Hamburg, and a sufferer comes by ship to Middlethorpe, he comes ashore, then it would be here..."

"Remember Aesop. We daren't shout 'wolf' too early," said Newcombe.

"Why don't you look out the delays between first cases and outbreaks starting in earnest in different parishes," Halliwell said to Leggate. "If we should do most good by distributing our bills when the first case appears, that's what we'll do."

"This assumes that we shall know of the first case," said Newcombe.

Newcombe made it his job to write to doctors in the area, informing them that cholera was now only a few hundred miles away, reminding them of principal symptoms, asking them to keep watch, requesting as a matter of urgency that they inform him if any case should appear.

Newcombe told Leggate that he had arranged with the secretary of the Middlethorpe Medical and Scientific Society for Leggate to read another paper.

"Just deliver something neutral," he said. "We need their support. This isn't the time to be fastidious."

Leggate felt the stab, but enlivened by his two friends' enthusiasm for his project, he agreed. "I'll give a historical paper about research in Magendie's laboratory. I'll call it: 'A memoir of research on the mechanism of cholera in the laboratory of A. Magendie.'"

"That won't cause trouble with Gadd, and I'll make sure that Anderson, Feather, and some others are there to speak up, if necessary."

Chapter Forty-seven

When Leggate stepped onto the gravel at Andervilliers Hall, as he and his wife responded to the invitation they had received from Lord Andervilliers, he saw the pillared entrance to a Palladian house far more grand than that of Evette's friend Madeleine. The Viscount himself, flanked by footmen, came forward to greet them.

He saw his wife walking towards His Lordship, who took her hand graciously. He experienced again that feeling which, whatever their troubles, allowed him to surmount them for her sake. She's recovered well, he thought, and the slight pallor that remains is somehow touching.

At dinner, seated to the left of Lady Andervilliers, Leggate would glance from time to time at his wife, who sat diagonally opposite him, at the other end of the table. She was only to be glimpsed by looking around the massive centrepiece of flowers and fruit.

So good she's better, he thought. And enjoying herself again.

"... the contrast between Battersby and Middlethorpe," Lady Andervilliers was saying, seeking to draw Leggate back into conversation. "I like the ships and bustle of Middlethorpe."

"And the new steamboats to Copenhagen," said Mrs Peake. "I've a cousin there whom I adore. We can visit now almost as if she were in the next county."

"It is very convenient," Her Ladyship agreed with Mrs Peake. "Do you make use of these modern means of travel, doctor?"

"Like you, ma'am," he replied, "I like the bustle of Middlethorpe. But having no cousin in Denmark," he smiled towards Mrs Peake, "I content myself with Yorkshire."

Her Ladyship glanced towards where Marian was sitting, then looked at Leggate, and said in the most friendly way, "I can imagine you'd be well content with staying at home."

"Your wife could be a musician in Paris or Vienna," whispered Lord

Andervilliers to Leggate during the applause after Marian's performance in the musical part of the evening. "If she were to study with one of the great teachers, I believe she could achieve extraordinary things. She has a wonderfully deft touch."

"She studied for seven months with Mendelssohn."

"Think what a twelvemonth would do!"

Leggate was delighted to pass Lord Andervilliers's compliment on to Marian, as the little audience rearranged itself while members of the quartet resumed their places.

Next evening there was a ball, and for Leggate its spoils can be briefly told. In the card room he stared at the gold sovereigns piled beside the players. There were one or two conversations. One was with a Dr and Mrs Morris. The doctor had, so his self-satisfied wife told him when Leggate invited her to dance, a very good practice in Battersby.

"He sees only the better people," she said. And then, after a short pause, "So you're from Middlethorpe."

"We are," said Leggate. "An interesting town, I think."

Leggate saw Mrs Morris's lips purse. "Interesting for its array of diseases, I suppose."

Leggate was introduced to several other people by his indefatigable host and hostess. He danced several times. But for the most part he found himself trying to look interested as he gazed at pirouettings that seemed incongruous among civilized people.

In bed that night in a large and luxurious bedroom, Leggate lay pensive as his wife splashed about in the adjoining dressing room.

We were so tentative at first, he thought, and the circumstance of her father's death, however should one know how to proceed in a case like that? And her mixture of pliancy and utter innocence. I felt I mustn't shock her. I'd thought myself experienced, but how could one know what to do in a circumstance like that? And then her announcement that she had heard from Caroline exactly what was to be done!

Leggate heard his wife still in the dressing room, and continued to muse. Andervilliers and his wife are very kind to us, and they seem to like us. Marian must be finished in there, whatever can she be doing? I was frightened of all that melancholy, bottomless it seemed, and not knowing what relation we had... Quite, quite different from Evette, one would think that those experiences in Paris were etched forever, they were so unmistakable, but they're no longer distinct now. But this... enthralling again, to be able with another human soul to have this effect, not just mental but physiological, with each other...

"What do you think?" said Marian, wearing her nightgown and appearing from the dressing room. "In an aristocratic bed I shall be Lady Camelia de Vere, and you, sir? An earl, a duke? Or perhaps the naughty stable-boy," she said, jumping onto the bed.

"You read too many romances."

"Better than you—you don't read anything at all."

Leggate steered away from this potentially troublesome subject. "I read the book of Nature," he said, undoing some buttons of his wife's nightgown, and turning a page of cambric.

"Do you think the Queen visits here, and Prince Albert?" she said. "Perhaps slept in this very bed, where they conceived a royal prince or princess."

"Are you conceiving, Your Highness?"

"I may be, I didn't know whether to tell you yet, or wait another month to see if it's sure."

Leggate felt moved by this strange mystery, that already in her womb, perhaps... "My dearest," he said.

"So then I'll be the compleat wife."

Leggate lay on his back, holding his wife's hand, contemplating the news.

"Why ever are you lying there," she said. "I said I suspected, not that I was certain. Don't you think we should make sure?"

Chapter Forty-eight

As he prepared his paper for the Medical Society, Leggate's articles came to a stop.

Back with my tail between my legs, he thought.

"Once you are in the Society, we can work with it, to meet the next epidemic," Newcombe had said. "They would be useful. It'll help get other doctors behind us. It's people's lives. You said yourself we must use every means. It's no time to stand on ceremony."

Newcombe has taken trouble, thought Leggate. And he's arranged for others to be there, personal considerations are out.

So Leggate delivered to the Medical Society his "Memoir of Magendie." He described the researches on cholera from sixteen years previously. He included the visit to Sunderland, the injections of mercury into blood vessels of the intestine of post-mortem patients. For controversy, he confined himself to two sentences in which he contrasted Magendie's theory of the virtual cessation of the heart with his own—as he put it—"more speculative idea" that the disease might be caused by perforating the blood vessels of the gut. He ended by saying, "This is a problem of pressing practical importance, not yet solved in a conclusive manner."

As before, there was polite applause from the small audience. As before, Gadd rose.

"Mr Chairman, Dr Leggate, gentlemen," said Gadd. "Ours is not a large society, and it is not usual for us to hear a paper with historical subject matter. On this occasion, the historical subject is one of the world's great physiologists . . ."

Leggate waited for the habitual sting at the end of Gadd's pronouncement.

If it's insulting, thought Leggate, I shall walk out, and there's an end to it.

But Gadd said, "It is an honour to hear from one who has worked in the master's laboratory. I welcome the paper."

Another man got up and asked how Magendie's theories on the physiological causes of cholera had progressed since the time of which Dr Leggate spoke. Leggate told him that the great man held still to his basic idea.

There were no more questions or comments. Next day, along with the letter from the secretary of the Society inviting Leggate to become a member, was a cordial note from Gadd inviting him to the Society's rooms next evening. "We have a pleasing ale that I can offer you," he wrote, "and we can send out for mutton chops."

When Leggate arrived at the Society's new Greek-façaded building, he found Gadd already there.

"Come in, my boy. Take a seat." Leggate had been shown into a well-proportioned room, laid out with mahogany chairs and tables. At one end, comfortable armchairs were arranged round a fireplace.

"Will you try some of our ale? Rowbottom, one of our chaps in the Society, takes an interest in Cobbs Brewery, you know, up by the waterworks."

"I know where you mean," said Leggate.

"Rowbottom has selected something he says is particularly fine, so we have them send a firkin over every few days."

"By all means," said Leggate, accommodating. He didn't drink beer.

"Get Dr Leggate a pint of our ale," said Gadd to the man who waited at table. "I'll have a glass of claret, myself. Then fetch us some mutton chops, will you."

He turned to Leggate. "There's not enough people to justify a cook, so we have to send out," he said. "Perhaps when the membership gets a bit larger."

Leggate thought it unnecessary to reply.

"You're still interested in the cholera? The Sanitary Committee is making progress in having nuisances cleared. Though whether it will do any good, I don't know. Like trying to clean the Augean Stables . . ." He laughed.

"I have been tracking the disease on the Continent," said Leggate. "It can't be long until it reaches England, if 1832 is anything to go by."

"You saw in *The Times* this morning?" said Gadd. "There's an outbreak in Hamburg."

"I hadn't noticed, I was out early . . ."

One hop across the sea, thought Leggate. How long now? Should I mention the Medical Society helping with making available water from Tafts Spring?

"So your practice is coming along?" asked Gadd.

"I divide my time between practice and research."

"I was thinking of your idea about contaminated water. It's not at all a bad idea, but too restrictive."

"I found extremely close correspondence between the cases in 1832 and drinking water from some sources, but not others, and there are practical implications."

"Quite so—there's a correspondence. But you said yourself, when you presented the preliminary version of your idea earlier in the year, you remember, that there was also correspondence with poverty."

"But less so. With water the correspondence is far closer."

"Still not exact."

"One wouldn't expect it to be exact."

"You studied in Edinburgh?" asked Gadd in apparent irrelevance.

"Yes."

"Me too. Best medical school in the land. Gives one a good start."

"I agree."

"People there think no end of Hume."

"The great Scottish philosopher."

"Exactly. You know what Hume said?"

"In what regard?"

"He said we can't ever see cause and effect directly. It is simply an idea, based on repeated experience. We experience repeatedly the conjunction of filth and disease, you said so yourself, that's the best idea we have. Now you say something else causes disease, but if so, it too must be an idea. It seems insufficient, just yet at least, to displace the experience of the many."

Leggate waited.

"I was thinking that if you carried your own logic a step further," Gadd continued, "one could imagine finding very close correspondence of death from cholera with something else, with drinking gin for instance. The poor consume oceans of gin."

Leggate was silent. This was a matter he had not allowed himself to think too closely about. Could there be some other agent than water, he thought, that has yet closer association with death from cholera?

"My idea, you see," continued Gadd, "not mine, of course, but the soundest that anyone can hold at the present, is that all these things contribute. Perhaps unwholesome water, as you say, though the water's a sight better than it was in 1832. But maybe gin too. Certainly idleness, and filth—especially filth. And debauchery. With the way those people live, there's no end of immorality. Our duty, as doctors, is to warn people against all of these. To concentrate on just one would be irresponsible, don't you think?"

"But the disease must have a physiological cause ..."

"No doubt it has, but you said in the paper you delivered the other day, I enjoyed it by the way, you said, did you not, that we don't quite know the physiological cause of cholera yet? Cessation of the action of the heart, increased permeability of the arteries in the gut ... just don't know. As well as the other theories. You probably don't think much of the theory that cholera is due to electricity in the atmosphere." Gadd laughed. "Neither do I. But you see what I mean? Not just cholera, but the fevers and other diseases. All follow the same laws; squalor and neglect cause disease."

"I agree, in a way ..."

"I know you do. So our medical duty is clear. Warn people against begetting children they can't afford, get them to avoid licentiousness, get the filth cleared up. Improve conditions generally, that sort of thing."

Leggate found himself unable to reply.

"So, you're married now," continued Gadd. "To the former Miss Brooks, charming girl."

"You know her?"

"Everyone knows her. Saw her at the hospital fête the other day, we're trying to raise more money, you know. Here's your beer at last. Whatever's taken you so long, Ernshaw?"

"I had to put another barrel up, sir."

"I hope you didn't joggle it about."

"Certainly not, sir."

"To our newest member: Your health." Gadd looked at Leggate, angling his head slightly forward to gaze across the top of his spectacles. "And congratulations on your marriage." He smiled, raised his glass. "Let me know what you think of that beer. It doesn't look too cloudy." He cast a fastidious glance towards Leggate's glass. "If there's the slightest bit of sediment from Ernshaw fidgeting the barrel about, we'll send it straight back, and you'll have a glass of claret."

"It's most acceptable, thank you," said Leggate. He raised his glass, mentally preparing himself to look appreciative at the sour-tasting liquid that would soon pass his lips. "Your health, sir."

Chapter Forty-nine

Marian had gone with a novel early to bed, to wait for her husband.

Gadd will be gracious, and John will feel welcomed, she thought. He's easily touched by kindness. And he'll come home and tell me that Gadd's not such a bad fellow after all, and I'll say he's condescending and likes to be deferred to, but not such a bad fellow, and John will feel relieved, and we can . . . what a curious activity this is, these marital rites, but how precious now, to think I never knew, didn't really even suspect when I would write all those words about friendship.

And so, lying in bed, instead of reading she thought about these rites, the effects they had on her, the imminent possibility of a further enactment.

Leggate made his way home quickly from the Medical Society. He wanted first of all to see in print the announcement of cholera in Hamburg.

Couldn't raise the idea of the Medical Society lending a hand, he thought, but Gadd seems to like me well enough.

At home, he went immediately to find the announcement in *The Times* that the cholera had reached Hamburg.

This'll be the chance, he thought. Years of preparation at the point of action. The water carts will allow the test of my theory, at the same time I'll try my new therapy, based on Latta's work in Edinburgh, which showed that rehydrating the patient by injections of saline fluid into the veins restored people. I might try that, but first I shall try another idea, try getting them to drink fluid continuously in small amounts.

He went to his study, opened his notebook to read again the idea of rehydration to make good the losses from the blood. Patients would have to drink the fluid in many small amounts, otherwise vomiting could undermine the effects, and perhaps make things worse. He had written a recipe for this fluid: spring water and common salt to make a saline mixture with a concentration similar to blood, but with sugar added to make it palatable.

"If the patient can drink enough of it and keep it down," he said out loud to himself, "pathological changes will be reversed, and so much simpler than trying to force it into a vein."

And histology, he thought, as he closed the notebook. Newcombe has an appointment at the infirmary. Together we can do microscopy on the intestines of post-mortem cases, to look for the planticles growing in the intestine.

He went to kiss Marian good night. She put her book down.

"Aren't you coming to bed?" she said, seeing he was about to leave the room again. "And what about Gadd? How did you get on?"

"It went quite well," he said. "Gadd's not such a bad fellow. He was quite pleasant, though he still thinks disease is every kind of filth all combined. I'll come to bed in a few minutes, just something I have to do first."

In his study he looked through his notes of preparations for the coming epidemic, but he did not concentrate well. The meeting with Gadd had discomfited him.

Gadd's idea of filth and depravity all together, he thought. In my article for *The Gazette*, that is how I would be attacked. The editor would write "Dr Leggate has found a correspondence with water. He acknowledges the correspondence with squalor and licentiousness. Though he says water is the cause, he fails to show why several causes may not contribute to each other."

Must find proof that it's the water alone, he thought. The water carts—those who take water from our carts will be protected, even though they live in the same filth. That's it. Afterwards when I visit homes of those who died, they won't have used our water. That will show the point, and with that the article will be complete. And Hume? Must read Hume.

When he finally returned to the bedroom, Marian was asleep. He was tired, and despite the sense of excitement he felt, he fell asleep quickly.

At breakfast, he said to Marian, "Cholera has reached Hamburg."

"When?" she said. "You didn't tell me."

"It was in *The Times* yesterday," feeling guilty, but not quite knowing why he had not told his wife this news when he had come in. "We have to make sure our preparations are complete."

She did not reply immediately. "Water carts and so forth," she said.

"Exactly... Gadd was quite affable when we met, though I had to drink some frightful beer."

"I'm glad he was friendly," she said. "You'll be thick with the Medical Society in no time."

"He said he'd seen you."

"He was interested to learn that we were married," she said.

"Did you go to the hospital fête so you could take him on one side for my application to the Medical Society..."

"I may have done," said Marian.

"What did you say to him?"

"It's not a matter of what I said. That's not how these things work."

"I wish you wouldn't," he said. "Science isn't something to be arranged."

"You won't let me help you write, or help with preparations, although the epidemic will be here at any time," she said with some warmth. "Can't you see, I don't want to be left out. I want to help."

Leggate did not answer.

"And Gadd could do you some good," she said.

"I know that," he said, thoroughly discomfited by his wife's show of umbrage. "I'm very sorry," he said in a tone that let her know he was aggrieved.

Marian reflected on this passage of words between them. He's a difficult case, she thought. When he behaves resentfully like that, I feel I'll let him get on with it himself, then he'd see.

To Caroline, later that day, she said, "It's difficult. He won't let me help him in ways that I could. He gets touchy. But I'd found out Gadd was to be at the hospital fête, and went and made myself agreeable to him. Then John reprimands me, can you believe it?"

"For oiling the wheels?"

"He won't go along with how things work. He doesn't realize that people in the Medical Society would be on his side if only he would take some trouble with them. He could get his discovery aired... and do you know? The cholera's in Hamburg, it was in the paper yesterday. John's making preparations. I could be very helpful there, but I'm still not wanted."

"In Hamburg," said Caroline in alarm. "Just a few miles away."

"I'm sure there's no need to worry. I'm sure John's got it right. If the disease does come, we'll follow his precautions on the handbill, you know the one I gave you."

Marian had been on the point of saying that she had been reading a lot, wondering whether there was some flaw in her husband's evidence, and that was why he was so morbidly sensitive about it, but this was something she would keep to herself. For her to mention to Caroline the slightest doubt would do nothing but cause an eruption of panic.

Caroline looked as if she were soothed by what Marian said about the precautions. "They're not always easy to manage," she said.

"What aren't?"

"Husbands."

Leggate, Halliwell, and Newcombe looked through their preparations in case, in a day or a week, the disease should leap across the North Sea. They decided there was nothing else they could usefully do until the first case occurred.

Leggate reflected on the possibility that it was not he who had found favour with Gadd but his wife. He felt ashamed.

While she's pregnant, he thought, I mustn't remonstrate. It's not good for her in her condition. I must let these things pass.

But Leggate could not let pass what Gadd had said to him. Although his evidence was an advance, Gadd had convinced him that the matter was not fully solved.

I showed a conjunction between water and deaths from cholera, he thought, but what if water does merely contribute to filth generally, if it only increases the effect of other filth, or what if something else were more closely linked? The Ericsons' farm where I first realized it, the kitchen floor was dirty of course, from the mud in the yard. But airy in comparison with the town.

Leggate stared at the ceiling. I know, he thought, I can show closer correspondence of deaths to water than to filth in the houses. When I do rounds after this epidemic, I'll estimate in each house the amount of filth, also of foetid air, and overcrowding which everyone says leads to depravity. And enquire about gin and suchlike.

While waiting anxiously for the disease to arrive, Leggate continued his thoughts and researches. He chose a poor area called Outham to enquire about looseness of the bowels, the usual summer diarrhoea, sometimes called "English Cholera," as if it were a relative of the real cholera. He found the condition commonplace; people saw it as without significance beyond the expulsion of some harmful matter.

He wrote in his notebook:

With warmer weather planticles increase, just as do larger plants. When taken by mouth, some sporules germinate and irritate the bowel, so a flux occurs, and sporules multiply and spread. Summer diarrhoea must be due to English planticles, here all the time; the cholera's a foreign species that only propagates here for a summer or two before it dies out.

As always Leggate had plenty of reading. He borrowed from Newcombe a copy of the first volume of Hume's *A Treatise of Human Nature*, in which causation, and correspondence, and the necessary connection that must exist between cause and effect, were very fully discussed.

Marian, almost as excited now as her husband at the near approach of the disease, would commandeer the newspapers in order to search urgently for news. "Look at this," she said, "in *The Times*."

"What is it?" asked Leggate. "Has it arrived in London?"

"No news yet, but there's this: 'The cholera is the best of all sanitary reformers,'" she read. "'It overlooks no mistake and pardons no oversight.' What do you think of that?"

"Very apt."

Marian knew to be careful. "Your handbill, and the precautions you and Newcombe and Halliwell are taking, are very thorough."

"The question is, are they thorough enough?" he said. "It's as if this writer knows what we're doing, and will be keeping an eye on us."

Marian, occupied by her pregnancy, was also occupied with her mother's move into her new house in Edward Street, and with her brother's acquisition of Hadby Grange, a large house with substantial land and a home farm, about eight miles away.

"Are you sure you'll be all right on your own?" Marian had continually asked her mother. "You could live with us, I've told you often enough, or even with your sister. For all you know Aunt Mitchell might like your company."

"You think we've not talked about it?" said her mother. "We have plenty of each other's company, which we enjoy, but in measured quantities. If you think loneliness will be a torment to me, you don't know me very well."

"No, I don't think that," said Marian, exasperated at her mother's obduracy. "It's not that, I'd just like you to be nearer."

"Quarter of a mile is not far."

"You, a quarter of a mile; William, eight miles."

"Why don't we drive over to Hadby with William tomorrow to help him see what he will need in the way of furniture?"

"Do you think he needs somewhere quite so large?"

"It gives him a sense of substance, and you're not to tease him about it."

Leggate oscillated between thinking the disease would arrive the next day, and thinking that somehow England might escape.

In the second week of September, Gadd invited him to join the Sanitary Committee. "You share our aims," he said, "to make the town safe from disease. Mears wants to retire from the Committee, never did much in any case. Should you like to join us?"

Had Gadd made the invitation from a sense that he had previously been unfair, or had he been charmed by Marian, or had he now taken a liking to Leggate? It was impossible to know.

But Leggate wanted to please Gadd. "I should be honoured, sir," he said.

"See you attend meetings, then. Mears was a lamentable attender."

Leggate saw that Gadd was in good humour. "I shall make it my first priority."

"Good. You have a special brief to attend to unwholesome water and drainage. That's your interest isn't it? But you must pull together, nothing hare-brained. If the cholera comes, which is almost certain, things could become very serious."

Chapter Fifty

For Leggate, it was almost a relief when, at the beginning of October, several seamen came ashore in Middlethorpe with cholera. It was the first landing of the new epidemic on English soil. Their ship had just come from Hamburg.

A telegram was sent to the General Board of Health in London, and the Board's two inspectors, Dr Sutherland and Mr Grainger, hurried to the town.

At a meeting with the Sanitary Committee, Dr Sutherland stated that the Prussian seamen had been exposed to the epidemic influence at Hamburg, and that on the voyage they had eaten plums being brought to market in Middlethorpe.

"The eating of a few plums," observed Dr Sutherland, "would certainly, under ordinary circumstances, have produced no such fatal results; but during an epidemic such indulgence is well known to be fraught with extreme danger."

"Plums," said Gadd to Leggate after the meeting. "They come down from London in order to tell us not to eat plums."

"It's nonsense," said Leggate later to Newcombe. "It's not some vague epidemic influence at Hamburg. I went to interview some of the sailors on that ship—the men were never ashore in Hamburg. The plums were from Hamburg though. They could have been rinsed with contaminated water, and so be covered in sporules."

The event was the signal for Leggate, Newcombe, and Halliwell to start distributing their handbills. The water carts began to circulate. Their drivers, employed by Halliwell, rang bells to attract attention.

"It's here!" Caroline wailed that evening, when Leggate told her the news, as she and her husband entertained the Leggates. "You said you'd advise me," she said to him. "I must make arrangements."

"You can be calm," said Leggate. "It's spread by water, or from any clothes a diseased person has soiled, from food that has been touched by anyone whose hands are contaminated. Have the people in your house all wash their hands before they eat or prepare food. Eat only cooked food, drink only spring water, and you'll be safe."

"You don't think we should leave town?"

"For where?" Leggate asked. "It's only affected a few sailors so far and, if it spreads, it will spread to other places. Take a handbill and read it to your servants, and make sure they get spring water from the carts that are coming round."

Leggate's speech had a calming influence on Caroline. "And I'll take some handbills to my parents' house," she said, "and my aunt's. You're sure this is the answer, not putrid air? And is there anything we should do in the house, to keep the disease off, limewashing or anything?"

"No limewashing, just the precautions I say, I think you'll stay well," he said. "And if you have any worries, you can ask me."

"Worries," said Caroline. "When it's a mile away, for goodness' sake. I feel beside myself with worry when I think my children will be exposed. It's not in the air, you say? What if you touch something inadvertently, a coin, a handrail, that one of them has touched—who knows who's touched what!"

"It doesn't work like that," said Leggate.

On the 6th of October a telegram summoned Dr Sutherland to Edinburgh where, as he later wrote in his report: "two cases of cholera had occurred simultaneously, one in an underground flat of a house at the top of Leith Walk, another in Leith in a wretched lodging house in a filthy cul-de-sac."

Leggate followed news of the Edinburgh epidemic in *The Times*. The British press was quick to broadcast the Board of Health's advice: cholera was common in low-lying areas, and was spread by foul air. *The Middlethorpe Examiner* added its voice. Drains must be cleared, decaying matter removed. Atmospheric impurity should be attacked in every way: expose bedding to the sun, dry-scrub homes without the use of water.

The Examiner also printed official advice from the Sanitary Committee. No one objected to the lines that Leggate added—his proudest public pronouncement yet.

Do not drink water with any matter suspended in it, or
that smells in any way unwholesome. Pure spring water
is available from pumps at Tafts Spring, and from six
carts which are circulating round the town.

Usually at first everyone denies an epidemic's existence. But in Middlethorpe, with the stricken sailors who came ashore and the visit from London
officials, the signs were so clear that people prepared for the worst. When
the water carts appeared, many ran out with pails. But they took other advice
as well. Some broke down partitions in their houses to improve ventilation,
others purchased a bottle or two of the many kinds of remedy that became
available, yet others wore red flannel belts around their bellies.

Almost all the citizens of Middlethorpe believed their chosen course was
effective because, although in Scotland the epidemic raged with a terrible fury,
and although Middlethorpe was the first place in Britain that the cholera had
touched, only a few isolated cases occurred in the town. In other parts of
England there were some outbreaks, but they were few, and the deaths were
not numerous. As with the English Chartist revolution of April that year,
the English visitation of cholera in 1848 was thought of as hardly more than
a fizzle.

In Middlethorpe, within three weeks of the sailors from Hamburg falling
ill, apprehension became complacency. Those who had listened for the bell
of the water carts found it more convenient to use the town water. Their
decision was safe. As to the post-mortem examinations that Newcombe and
Leggate had planned, only one case reached the infirmary. They could not
get near it.

Leggate, however, was intensely occupied. He felt excited to be working
with the Sanitary Committee. He visited every part of the town, attended
every doubtful case he heard of. He thought to travel to Edinburgh to try
his hydration treatment. But with Marian expecting their child, although
confinement was not due until April, he believed he should stay by her side.

For a month Marian saw little of her husband. She did not complain, for
she knew this was his most important moment, and when the epidemic failed

to develop, the coming child, a source of gladness for them both, drew Leggate back towards his wife. Though she could not say so, Marian was utterly relieved. She was glad to bring his mind back within the household.

"I think the blue room would be best for a nursery," she said.

"Shall I make some drawings to be framed, for the room?" he said. "Or you could do some water-colours."

A month after the seamen from Hamburg had fallen ill, Caroline's alarm about cholera had also stilled.

"Caroline and Peter want us to play whist with them," Marian said to her husband. "Once a week perhaps."

"Very well."

"Caroline says it's a framework for conversation. She says her husband can talk endlessly about goings-on in the town, but that he's not good in mixed company. So would you mind whist?"

At the card table, indeed, Peter Struthers would chatter happily about the fall of the cards, and indulge an inoffensive flirtation with Marian.

Though Leggate could think about the coming child, his only other mental space, it seemed, was occupied with the cholera. At cards he would speak largely when spoken to.

"I wish you'd concentrate," said Marian after the third of their whist evenings.

"I do concentrate, I'm not very good at cards, that's all."

Leggate was indeed the most indifferent player in the group. He did know to lead the smallest card of a long suit. When he had to respond to his partner's lead, he would usually remember to play out his highest card in that suit. But these were almost the only rules he seemed to master. He seemed never good at managing trumps, he missed signals that his partner's cards were intended to give.

It seemed odd to Marian that an intelligent man should be so inept in this pursuit, but Caroline and her husband did not seem to mind. Indeed, Peter Struthers, when partnering him, liked saying at the end of a hand: "Good Heavens, man, didn't you know I had the Queen of Hearts?"

Leggate would reply: "I'm sorry. I guessed that Caroline would have it."

So long as it doesn't tell us about ourselves, Marian reflected silently after one of these evenings. With a partner you communicate, she thought, and make the most of what you hold between you, for if not, deficiencies can't be compensated by the other, and even a good card can be trumped.

Chapter Fifty-one

Marian saw in *The Examiner*, in November of that year, that Warrinder had taken to writing about the cholera.

> In Middlethorpe the disease did not take hold this year
> as it did in 1832. If we ask why, the answer is not far to
> seek.

Warrinder described improvements in the town in the previous sixteen years. He mentioned the new waterworks, improved drainage, superior new houses. Warrinder mentioned, too, the clearance of nuisances begun by the Sanitary Committee. He concluded:

> Some doubters have sneered at the benefits of modern
> progress, but when progress is measured by several hun-
> dred people, working breadwinners, as well as mothers
> and children, being alive now, who would have perished
> from cholera were it not for these improvements, who can
> doubt that the strides we have taken are for the good?

Could it be that we really are safe now? Marian thought. She had been reading about cholera in her husband's medical books, and had become more knowledgeable than most doctors. She wondered whether to ask her husband if Warrinder might inadvertently have hit on anything, but decided she would not.

Leggate saw the article. It made him angry. Everyone can see the town's still strewn with rubbish, he thought, but such a mind as his doesn't regard anything evidential. How very fortunate to be so sure of oneself, how good to bask in such warm self-congratulation.

Although his talk with Gadd, awaiting beer and chops at the Medical Society's rooms, had impeded him once more in sending off his articles to *The Medical Gazette*, he continued to rewrite his manuscripts. At night, if Marian had gone to bed, he would doze over Hume's treatise. It was written with such unfussy good sense that it was impossible not to be attracted, yet the philosopher drew conclusions that, where Leggate thought he understood them clearly, still seemed hard to apply to his own science. But they spurred his thinking. If the epidemic should come again in force, he considered how to use it to prove his case against all doubters.

But the failure of any epidemic to materialize in Middlethorpe plunged Leggate into self-doubt—perhaps Gadd was right, he thought, and water's no more than a small contribution.

But then a surge of annoyance: what arrant nonsense Warrinder gets away with, he thought. Was Gadd behind that article? He won't attribute causal power to water, but he's happy enough to tell Warrinder that the town's escape is due to removal of nuisances—a pile of rubbish is shovelled up on Tuesday, hurray we're all safe, then a further pile of rubbish is in the same place on Wednesday.

Leggate turned to his notes, looked over what he had written concerning the handful of cases that had occurred.

I'll speak to Gadd, he thought. He'll say: "You're quite right, my boy; shouldn't have let that newspaper chap write what he did"—but what good would that do?

Leggate's thoughts gave way to a more persistent concern: how could he use the few sporadic cases that had occurred after the stricken seamen had come ashore? He had visited families, gathered all the information he could. Two cases were doubtful at best. He performed some arithmetic, always soothing. He divided the number of cases by the population of the town, to find proportions. First including the sailors, then excluding them, then excluding the doubtful cases. It made no difference. As compared with the those who had died in 1832, the proportions were tiny, of no significance.

Then a realization struck. "Stupid, stupid," he said aloud to himself.

The brain of an animalcule, he thought. I'm still using my method for 1832. But now it could not have worked.

Leggate lay on his couch, heart pounding. He had imagined a large epidemic, he had planned afterwards to visit the house of every dead person, enquire whether that person had drunk town water, or only water supplied from Tafts Spring. Now he saw that if the disease had broken out, this method could not have proved his idea.

Why did I not think? he said to himself. I should have been taking addresses of all those who used the water carts, and bound each one, on his oath, to use no other water. Asking retrospectively, it would have been all but impossible to find who'd drunk only spring water because they would not have died, so how would I have found them afterwards to calculate the proportion of those who stayed healthy? It's a lucky escape, for if the epidemic had come, I would have wasted the opportunity.

He rose, paced the room. Then he lay down again. Must calculate the rate of disease among those who drink my water, he thought, compare it with the rate of disease among those who drink town water. Comparison. Must be comparison! That's how to beat Gadd and Hume.

He scribbled agitated notes:

Not even a problem that is new to me, I'm just not thinking properly. It is the same problem as at the Hôtel-Dieu comparing the pneumonia sufferers who had blood taken, and those who had not. Here's the right way: one hundred take spring water, perhaps one dies—one hundred people take town water, ten die. Compare the two—one compared with ten is perfectly convincing, but to do the sum, must keep track of every single one who takes the spring water—studying the disease is not enough, it needs the comparison.

Leggate found himself imagining that he was watched over by some special providence, that the disease's visitation in the autumn of 1848 had been a warning to him that he had not yet been ready.

It's like the previous epidemic, he thought. That started in 1831 in Sunderland and some other places, then went into abeyance in the winter, and erupted again the next summer. The real epidemic will come next summer.

He felt guilty at wishing it on the town, but so caught up was he in belated realization of the evidence he needed to prove his case that he continued writing.

TEST OF COMPARISON
Must not only work backwards from those who die, must compare forwards: proportions who *do and do not take the disease* while drinking only the spring water, with same proportions of those drinking the town water.

To assay hydration therapy is the same: one hundred cases who take my hydration mixture, compare with one hundred cases of another physician, ask him to give the usual remedies, calomel, opium, whatever he likes, but record the cases carefully. I need to arrange in advance.

Excited though he was by these thoughts, he kept them close. The handbills had been no more than a flutter of anxiety, the water carts unnecessary. Ashamed of his confident predictions of the epidemic, he did not tell his new plans for comparison to Halliwell or to Newcombe, let alone to his wife.

Alongside the excitement of his new idea, he suffered continual anxiety. We can't be certain whether it will come again, he thought. Perhaps the 1832 epidemic was just a single severe episode, and the disease has lost its force in England. Perhaps Warrinder's right that conditions are cleaner, so the soil's no longer fertile for the sporules' propagation here. All that thought, all those preparations ... all for nothing, and not even a published paper to show for it.

Marian became concerned about her husband's moodiness and preoccupation. In her eyes the excitement he had shown in the weeks after the illness of the German sailors had turned to a dull withdrawal. By January his spirits were low.

It's hard to account for, she thought. He's thinking, thinking all the time, but dejected, he must see some flaw, perhaps wondering if sanitary improvements have prevented the disease from taking hold, so making all his research irrelevant.

But there's something underneath all that, some underlying pattern between us. When I was melancholy, he could be strong. Now I'm feeling full of life, with the baby, he's in the dumps. Does he need some medical emergency to be happy?

She regarded the aneroid barometer that hung on her wall. Instead of a pointer its spindle was mounted vertically to hold a light beam, on one end of which was a pasteboard cut-out gentleman with umbrella, on the other end a lady with parasol. When atmospheric pressure was low the gentleman emerged from a doorway on the left of the toy house in which this contraption was mounted. When pressure was high, the gentleman stayed indoors and the lady with her parasol would be out.

Is that what we're like? she thought. One in and the other out, one down the other up?

Surely not. She sighed audibly. Surely it can't be like that. We managed for a while when I got better, after we'd returned to Middlethorpe. There was a time of relief for us both, but so short. It's his science, that's what's important to him, not me at all. And when it's not going right, it distresses him. He can't bear even to talk to me about it. And he seems no nearer to sending off his paper.

Marian's girth was now substantial. She said to Caroline, "I feel quite healthful, but at a loose end, waiting for the baby, I suppose. And John's so withdrawn, moody—I can't stand it."

"You'll have a good deal to think about before long," Caroline said.

"We used to read to each other in the evenings, but he's either writing and rewriting, or off to meetings of the Sanitary Committee, or the Medical Society," she said. "I wish I'd never seen Gadd for him."

"It won't be long now," said Caroline. "The new baby will change everything."

"I believe he's trying to think what would be conclusive evidence," Marian said—she avoided saying that she knew her husband thought the real epidemic might come next summer.

"If only he'd talk to me about it," Marian continued. "I can think; we could think about it together. The last I really knew about where his thoughts had reached was the excitement with the handbills and water carts."

Chapter Fifty-two

Left to herself, Marian still read a good deal: philosophy, and more medical books, including now the subject of childbirth, and nursing a newborn.

When the human seed is planted, she thought, we leave behind the power of human will. Whether the seed takes is chance. The course any new life takes is chance, and chance again, will the child be boy or girl, healthy, misshapen? Will all go well? The confinement? The delivery? Afterwards, will the child survive? Will the mother?

When she spoke to Caroline, Marian asked, "Does it hurt a great deal?"

"My first labour was long," Caroline said. "But when the baby is there, it's forgotten, there is too much else to think of."

"Women die from it. There's nothing one can do."

"You couldn't be in better hands. Your husband—best if Dr Newcombe delivers you, whom you have known all your life, then John can concentrate on being with you."

That evening, in bed, Marian spoke to her husband. "The baby is kicking again. Does that mean it's all right?"

"It's flexing its limbs, for the outside world."

"But healthy, the child will be healthy?"

"Let us pray so."

"What if I should die?"

He held her. "I have been thinking about the birth too," he said. "Women don't die from the delivery, but afterwards from a fever they contract while being delivered, but which takes hold later. I believe there are sporules, not the cholera ones but others, which enter when the body is open, and cause the fever. I've always washed my hands thoroughly when attending women in labour. I've had very few such fevers. The handwashing was a habit. Before now I did not understand the reason. Now I do."

"Will you deliver me, John, or Dr Newcombe?"

"We'll ask Newcombe, that'll be best. I can hold your hand and talk to you during the labour."

"We will make it safe," she announced.

Thinking of her plans, she said, "I shall have our spare room for my delivery, have it cleaned and scrubbed, fresh linen, and two changes of linen in case I shall be in labour for long, and shall need them. We'll close the room up, to keep it clear of sporules. You tell me everything else we'll need, towels, water, anything. I shall have them prepared."

Two months before her confinement was due, as sometimes happens in the last part of pregnancy, Marian felt an inner glow of health. She began as well to find once more with her husband the harmony for which she had longed.

"I feel a quite lovely serenity," she told Caroline. "And John seems no longer so tortured by his research, he's more attentive again. Perhaps we'll just settle down and be nothing extraordinary, the doctor and his wife, with their children."

The labour was only fifteen hours, the birth less difficult than many. Leggate sat by Marian's head, held her hand, spoke to her through the ordeal, while Newcombe attended to the delivery. The room was equipped, scrubbed, spotless. The baby was born, a boy. Marian kept free of infection. The baby was christened Frederic. Caroline and Peter Struthers became godparents.

Leggate had stopped ruminating on the cholera's return. The pregnancy, his wife's health, the new child occupied him.

The first few weeks of a child's life are a tender time for parents. Marian found herself affectionate, watchful, contented. Leggate would gaze at his wife as she suckled the child, and was moved: the infant so intimately connected, the mother abundant. He discovered a ready affection for his new son. Wife and husband could devote themselves, become mother and father. Was this the ingredient they had needed?

Chapter Fifty-three

Leggate continued, in the background of new concerns, to keep an eye open for announcements of the cholera, and there were some: smallish outbreaks here and there throughout the country. But as the spring of 1849 progressed, new outbreaks became fewer. Frederic's birth coincided with what seemed to be the final departure of the disease from England.

Perhaps, as Warrinder and many others suspected, sanitary improvements in England had been sufficient to prevent the disease taking the strong hold it had exerted in 1832, while the rage of the disease in Scotland showed, perhaps, that there sanitary progress had been less complete.

But with the baby not much more than a month old, the idea that England had escaped was shown to be wrong: fierce new outbreaks of cholera occurred in Durham, in Liverpool, in Gloucester.

Leggate considered that now he and Marian had regained the intimacy of the previous summer, and now wife and baby were safe, his attention within the family was no longer urgent.

The disease has been lurking in cesspools and drains during the winter, he thought, now it's breaking out again, barely a hundred miles away. It needs merely to get into our river, and it will be all over town. We're not finished with it, a chance to beat Gadd, and the miasmatists, and the contagionists—all of them.

He read through the plan formed in the autumn, to put his theory to the critical test of comparison. He was eager now to try out his idea on Halliwell.

"I intend an exacting trial of my theory," he said, "otherwise people will continue to think cholera is just a combination of depravity and filth, with water playing no more than a minor part. We'll be in servitude for ever."

"A new scheme?"

"Cholera is in Durham, and may be here again soon. In 1831 it became quiescent in winter but then returned in the summer of 1832, as one might expect if it depends on tiny sporules."

"And break out again here?"

"Here's my scheme. I shall persuade people living in some lodging houses, places that would be expected to succumb, to drink only our spring water if the epidemic strikes. I shall keep track of everyone who follows the advice. But here's the thing: those I can't persuade, they'll be more important even than those I can, for they will provide my comparison."

Halliwell listened to this outpouring.

"In my practice," Leggate continued, "I've some women who are prostitutes—you know that—I shall instruct them. When the disease strikes, I'll make sure they drink only the spring water, and they'll be protected, and I'll compare them to those who drink town water. Then nobody will be able to say the disease is just a matter of depravity. I shall find other lodging houses that are the poorest, and do the same thing, then no one can say it is a matter of squalor or dirt. Then I can write the article for *The Medical Gazette*. I'll have evidence that even the most dissolute and the poorest can be exempted from the disease."

Halliwell hesitated. "It's certainly bold. You supply clean water to some houses but not others. All else will be similar. Is that right?"

"Then only the water can be responsible. The argument that some generally prevailing vice or filth is to blame will be scotched. Is my reasoning correct?"

"Give me a moment to think . . ."

"I can make other enquiries, to take care of the alternative ideas, drunkenness, licentiousness, stale air, to show how none of these is closely associated . . . If I am right, that is."

"Those who don't take the spring water? Isn't it wrong to withhold it?"

"I won't withhold anything."

"Only if they outright refuse . . ."

"Or if they don't comply, about which I shall enquire subsequently. If they do take town water then fall ill, I shall give my rehydration treatment. I won't give them up."

"Shall you tell the Sanitary Committee?"

"Not until after the event. I shan't want Gadd undermining me, or quibbling. Afterwards, I shall address the Medical Society at the same time as I send the paper to *The London Medical Gazette*."

"What if the people in your lodging houses survive, as you suppose, but someone were to say, 'He just chose houses that were clean and well kept.'"

"I'd say, 'Not at all. Some were of the lowest kind.'"

"You might not be believed…here's how you could do it." Halliwell, having caught the excitement of his friend's idea, now entered into it. "You could choose pairs of houses, equally filthy. Write the address of each house on a card. Then put each pair into a hat, shake them up, and ask someone trustworthy, who has no interest in the matter, the mayor or someone, to pick the cards out of the hat, the first house gets spring water, the second does not. Then do it for another pair, and so on."

"You are a purist," said Leggate. "You wouldn't suggest it if you saw the people. If the house of anyone I know didn't come up, I shouldn't be able to carry it through."

"Just thinking aloud."

"I shall find it difficult in any event. I'll try to persuade them. Once I know of a house, if I do not exert myself to the utmost … I should like to save the whole town." He paused. "The world, really, but first the town."

Leggate went to see Amy that afternoon. She was playing with little Florence, while talking to the housekeeper in the kitchen.

The housekeeper got up when Leggate entered. "Shall I take the child for a bit, while you talk to the doctor?" she asked.

"Just for a few minutes."

The housekeeper left the room with the child. Leggate paced up and down. "Amy," he said, embarrassed, "may I ask you a kindness?"

"Of course, sir, anything. You don't know how grateful we are to you."

"It's complicated to explain …"

"About your marriage? You can talk to me." She took a step towards him. She touched his wrist.

Leggate was shocked, but moved by the affection in her touch, he squeezed her hand gently, then let it go.

"No, nothing like that. It's the cholera. It may visit the town again this summer, hundreds of people could die. I think I've discovered how it's caused. You saw the handbill that went out?"

"I did, and we took the water from your carts . . . least till it was safe."

"The disease comes as a kind of poison." Leggate did not describe animalcules and sporules. "It gets into the water—when people drink it they get terrible diarrhoea and die. But I cannot convince people that this is how the disease works. They think it's from filth, overcrowding, vice, and so forth. So, if the disease comes again, which it may, I've planned an experiment. I'd like all the women in your house only to take water from our carts. Perhaps for two or three months, if the disease comes again. I'll arrange for the water to be delivered. We can give you pails."

"It's no hardship to us if it's delivered. But this is you doing us a kindness. What was it you wanted me for, sir?"

"You know some of the street women, you have spoken to me about them. I would like to try and save some of them as well. Could you take me to some of their houses, help me explain, so that they trust me, and would do what I say. I want to show that if the disease comes again, people who drink only the spring water, even if they're poor, or disreputable, won't die."

Amy considered. "You need to convince others, other doctors?"

"You have it."

"When do we begin? This afternoon? In the evening the women are likely to be out."

So Leggate accompanied Amy.

"You don't mind being seen walking with me?" she said. "I wouldn't want to be the cause of scandal."

"My wife quite understands this work is important. I suppose it might not be so good if my wife's relatives saw us together, but they would scarcely be in this part of town."

In the next weeks they visited lodging houses of women whom Amy knew, and Leggate carefully compiled lists of names. Each house needed several visits. He got the cooperation of two other houses, to whom he promised to deliver water each day. With the house in Friars Walk, a total of twenty-nine people agreed to take the spring water. The keepers of four other houses would have nothing to do with it. One woman who kept a house did not trust Leggate, three others said it was too much trouble when water was piped straight indoors, and in any case they had been safe enough in the last outbreak of the disease, but they allowed him nonetheless to take the names of those who lived there.

During their visits Leggate talked with Amy. Her life was without many comforts, yet she retained an independence of spirit. "I have a gentleman," she said. "Likes me not to go with anyone else. He gives me enough to dress nicely, and live on, just about. To him it's nothing. So I don't have to be in service or work in a factory. If one of the other girls is ill, or has her month-lies, or if I need to earn a bit extra, he needn't know.

"I'd like to find a way of doing better for myself," she said. "Connie, and Liza, and me don't want this for little Florence. I got a schooling you know. I can read and write. We'll make sure the little one gets school, but she should do better, not live in this way. I used to think about getting married. I could make a nice wife for, I don't know, a clerk or someone with a bit more money than a working man. But now we've all agreed to bring up the baby, that makes it more difficult."

On another occasion: "That article that was in the newspaper, did you read it, sir?" Amy asked. "In *The London Times* it was. My gentleman left it for me, wanted to know what I thought, if you please. About fallen women. I suppose he thought I'd have an interest. 'Plain ridiculous,' I told him. The newspaper said the country's slipping into immorality and vice. The article said that when respectable women is deceived by men, they fall helplessly into the gutter.

"I don't believe it. I know quite a few in the town, and I don't know a sin-gle one like that—perhaps there's some but I never met one. Mostly they are like me—we're not fallen. We was never anything better. We're on our way up from something a sight worse. We'd like to be better, and that's the truth. I can't think why those newspapers have got it so upside down. I mean they're supposed to be educated, the ones who write the papers, aren't they?"

"You should tell Warrinder. He's in charge of the Middlethorpe paper, he's the editor now. He still visits your house, doesn't he?"

"I wouldn't talk to him. He's too stuck up."

"You could write a letter to *The Times*. It might do some good. I would help you with it, if you needed."

Whatever am I saying? he thought. Am I to give lessons on writing to the newspapers?

Leggate was unfortunate enough, one afternoon after an outing with Amy, to encounter Warrinder at the house in Friars Walk.

Leggate could not resist making a gibe: "I thought your man was properly hanged, at least that's what you told us. Or is there another case in this very house?"

"I could ask what you are doing here, yourself," said Warrinder.

"If you wish to know, I'm trying to warn people against the dangers of the town's water supply. You refused to print my letter, so I have to go from house to house."

He heard his own words. In one ill-considered sentence he had made plain his chagrin while sounding absurd.

"Indeed." Warrinder arched his eyebrow. "Perhaps you should thank me. Perhaps in such visits you get a deal more satisfaction than from any letter in a newspaper."

"You should use your paper to do something worthwhile, instead of wilfully obstructing the truth."

On Warrinder's face he saw an expression of profound dislike. I wonder if he sees the same expression on mine, he thought.

Leggate left the house, too angry even to say "Good day."

Leggate went also to lodging houses of the lowest type, each of which had been visited by the cholera in 1832. The first was a large house on four floors that had once been home to some wealthy family. Fifty-three people lived in it now, beds in every room, on the landings, in the cellar. People crowded together, both sexes, all ages, often three in a single bed. Some beds vacated by day workers were used by those who worked at night. Since there was no privy, the most bestial matter accumulated. Fireplaces were closed by boards to keep out draughts, windows were shut, so the whole was sealed. To one who did not live there, the stench was insupportable, yet families and individuals had taken up residence, made the place their home. Leggate prevailed upon the keeper to take water from the carts which would visit twice a day.

Next day he took a quantity of jugs and buckets. Leggate spoke to every adult in the house, enjoined them to use only his water. He determined to visit every second day if the epidemic came, to ensure his instructions were followed. He enlisted support in four similar houses. In five others the keepers were uncooperative. Two thought it would be too much trouble; they had water piped to the house. One had a standpipe directly outside. In the other the proprietor thought Leggate an official of the Town Council. He nevertheless asked permission in these houses to offer medical aid to any who contracted the disease.

After a month of arrangements, Leggate looked over his lists, and considered his numbers. Two hundred and seventy people would take spring water, three hundred and twenty-three would not.

As Leggate and Amy returned towards Friars Walk after visiting the last of the houses to which Amy had introduced him, she unaffectedly took Leggate's arm. They walked together like a respectable couple. Leggate accepted the companionable gesture, gave her hand an acknowledging squeeze with his forearm. As they walked, she held his upper arm against her breast, unobtrusively. It was not accidental. He took it as a mark of affection, creating a conspiratorial frisson.

"So you've got to go," she said. "I can't offer you tea?"

"Thank you. I must go. I'll see you tomorrow."

Marian, recovered completely from the birth, considered her situation as a mother. In love with the baby, pleased with the new nursemaid, she was contented with everything except her husband, who was now preoccupied, off again on the island of science a hundred miles away.

"Must you be out so much? I don't see you all day or all evening, and you don't see Frederic at all." Marian tried not to remonstrate.

"I need to make this last push, to prove my theory."

"If only you would talk to me about it. You know I could help if you'd let me."

"Don't scold," he said. "It's the last push, I promise."

"Don't scold!" Marian reiterated, and heard her voice harsh, as if the scolding had been there, waiting to burst out. "How can you say I'm scolding? I'm offering to help."

Leggate looked at her, but did not reply.

Marian read disdain on his face. As if he's already left me again, she thought, like last autumn with the cholera scare, and then during the winter.

"Can't you even reply, when I speak to you?" she said.

"If the disease comes again this summer or autumn, I must be prepared. You married a scientist, not an indolent landowner."

Marian could not believe the coldness in his eyes, could not ever remember him having spoken to her so sharply.

"Then Frederic and I will manage without you," she said, and left the room.

Marian received visits from Caroline. "I've been deserted," she told her friend. "From being companionable for a month or two—a true friend it almost seemed—I never see him."

"His research?"

"I feel so frustrated. He won't talk about it, just says he must be completely prepared."

"Because he thinks it's coming again?" said Caroline, her eyes full of alarm.

"I should tell Mrs Huggins that she should find another situation," said Marian. "I have all the right qualifications for a housekeeper now, arranging the house, making sure the right food's got in, so we'd be able to save her wages."

Caroline was so shocked at Marian's sarcasm that she forgot her alarm at the idea that the cholera might return.

"We're incapable of living like a normal couple," said Marian. "First I am so ill with that frightful affliction, so that I'm just another mad female patient—though I had his attention at least. But I learn from my errors, and I come to see and even accept that marriage cannot be an equality, but that the woman has to make adjustments, and I make them. I make him a home. I do everything a wife should do—then he's not even there!"

"Men get involved in things," said Caroline.

"Perhaps I should be ill again, then he'd notice me."

Caroline did not reply.

"With friendship," Marian said, "when I used to write about it, I was drawn to it because somehow I knew that one could make choices, and there could be a kind of equanimity in it. But with love, and marriage, there are such unseen forces that one doesn't guess at them, and can't even begin to contain them. At least I can't."

"You sound so vehement."

"I can't even say anything, it only drives him farther away."

"You frighten me, when you're like this," said Caroline.

There was a long silence during which a thought flooded Marian's mind, a doubt, an anxiety—what if her husband's theory were not correct? But this was not something she would voice to Caroline.

"I want him with us," she said at last. "And whatever good is all this research, if no one will ever read it?"

Marian, despite her promise to herself, said to her husband again: "You're sure I can't help?"

"I've got a scheme, and, if it works, the paper will write itself in a jiffy."

"Why won't you tell me? You don't even talk to me now."

"I will tell you about it. It's just not been ready, so that I need to keep turning it over in my mind, until it comes right, there's nothing much to say until it is right, I still feel in a muddle—you don't want to hear about that... you remember when you were in a muddle?"

Marian was silent. I told you all my muddle, she thought. All of it . . . though I suppose not immediately.

She saw her husband realizing that she was not going to reply, and that he needed to say something else at least.

"With this I think it'll come right," he said. "I've arranged a test of my idea, to supply pure spring water to some of the lowest lodging houses in the town, to find if it protects them, as compared with other houses, equally low, who won't accept the spring water. It's an experiment. Either it will make sure of it, or show my idea is wrong."

Again she did not speak.

"If it's wrong I'll give up research, just be a doctor, and husband, and father," he said.

If it's wrong! she thought. If it's wrong you could talk to me and we could see where it was wrong, and make it right.

Instead she said, "You promise you'll send the paper off without worrying whether anyone can argue with it?"

"No one will be able to argue."

She went up to him, pressed herself against him. "I'd like you back," she said.

Leggate, though he had been confident in his reply to Marian, received a sudden uprush of worry about his scheme. His answers to his wife echoed in his ears with the sound of hubris.

It's one thing to draw a conclusion, he thought, and write it down, even to have it published, but this is the more severe test. What if the epidemic comes, and I am wrong? What if I tell people in the lodging houses they will be safe, and they are not? All those people! I'd be responsible for their deaths. Perhaps they'd be better doing something else, perhaps leaving the town. Perhaps I shouldn't be meddling in it.

The more he thought of the test he planned of his theory against the most stringent of criteria, whether people lived or died, the more agitated he became. He knew that from being solicitous for him, Marian had become exasperated. It was as if not he, but some wooden other man, inhabited the house, neither able to speak spontaneously nor answer his wife satisfactorily.

What if there's a flaw? he thought. Not just a flaw to be seen by professors of medicine, but a real flaw. Others have been convinced they had the answer and were wrong. Magendie thought it. Why not I? Why should I be right? What was it that Marian read me from *The Times*? "Cholera is the best of all sanitary reformers. It overlooks no mistake and pardons no oversight..." A murderer such as Biggs kills just one unfortunate. If I am mistaken, I could have a hundred deaths on my hands.

Chapter Fifty-four

On the 23rd of July Leggate wrote in his diary that the cholera again erupted in Middlethorpe. His first thought on hearing the news was of Warrinder's article on the efficacy of the town's sanitary arrangements. So much for complacency, he thought. Not just specious, he's downright dangerous.

In Leggate's next thought, the sword of apprehension thrust into his own bowel: here, now, he would be put to the test.

Within two days of the outbreak Dr Sutherland of the Board of Health was again in the town. He arranged to be taken round, to make rapid inspections. A combined meeting was called of the guardians of Middlethorpe and surrounding parishes, together with the Sanitary Committee of the Medical Society.

Dr Gadd took the chair. He introduced Dr Sutherland, who started by pronouncing his satisfaction that, since his last visit, a number of nuisances had been abated.

Then he continued, "In other respects the protective measures I have seen are few. No houses of relief have been established. Great piles of filth are still to be seen everywhere. Yesterday I observed an area called Outham, a triangular piece of land of about two acres. There, night soil and other manure is deposited. It is surrounded on all sides by dwellings. People there accumulate this matter to sell for agricultural purposes. It lies close to doorways, heaped against walls and under windows. The people even consider it wholesome rather than otherwise." Dr Sutherland was agitated.

Sutherland's quite right, thought Leggate. It's a disgrace—the area was the one he had visited to investigate summer diarrhoea.

Sutherland became more calm. "I am informed that in the entire town no houses have been limewashed. Dispensaries are insufficient. No house-to-house visitation by medical officers has been adopted. I cannot but conclude that there's a very great deal to do."

Men sat still, like schoolboys being reprimanded.

Sutherland went on to give details of the measures to be taken, giving special attention to visitation. "It is now established," he said, "that premonitory diarrhoea is part and parcel of cholera. This fact has hitherto been neglected because, without specific questioning by a medical visitor, whether through modesty or for other reasons, patients do not voluntarily speak of looseness of their bowels. The poor reject the expense of medicine. But when medical visitors have enquired carefully, wherever the cholera has started to prevail, many cases of diarrhoea have been discovered, even in the winter when such affliction is normally rare. If untreated the condition can pass within two weeks into the cholera itself. Once this has occurred all remedies are ineffective. This is of the first importance—if premonitory diarrhoea is treated in its first few hours and treatment given, the patient is saved."

He gave the example of Dumfries. "Medical visitation began on the 10th of December," he said. "At first with only one visitor, but by the 13th there were twelve visitors, and from the 18th onwards, fourteen.

"At the beginning of visitation, on the 10th of December, there were thirty-seven cases of full cholera, on the next day, thirty-eight. But under treatment the cases of diarrhoea were prevented from passing into cholera. After ten days, with visitors at first treating more than forty cases of premonitory diarrhoea every day, cases of cholera fell to none. Thereafter, they never rose above three cases a day." Sutherland paused.

"I prepared a diagram for the Board of Health. I made a copy which I now pass round. The red line shows cholera cases. You see them falling away after visitation began. The black line shows premonitory cases, which at first are high, then they also subside as the epidemic is abated."

He passed the diagram to Gadd, who inspected it for a length of time before passing it on.

Sutherland continued, "The evidence, gentlemen, in the instance of Dumfries, as to cause and effect is as strong as any such case admits of. It is, I think, perfectly conclusive. During this winter and spring, comparable results have been found in other towns. Mr Grainger, Superintending Inspector for London, attests to a precisely similar experience there.

"I know Dr Gadd and the Sanitary Committee here are in complete agreement with the aims and ideals of the Board of Health. We all agree it is a law of nature that sanitary arrangements adduce to health, whereas filth predisposes to disease. I cannot urge on you strongly enough that you should proceed by every available means: abatement of nuisances, every measure to purify the atmosphere, provision of healthful houses of refuge, and, above all, visitation. Do not rely on one measure alone. I shall leave a list of things to be done. Use the full force of the law to require removal of nuisances, and, above all, appoint medical visitors with haste."

Gadd rose to face the Board of Health's superintending inspector. "We thank you, sir, for your presence here, your timely advice, and your account of the very interesting new theory of premonitory diarrhoea. I should like to ask a question."

Gadd had a sense of the theatrical. He paused for longer than necessary. "You have told us of the impressive results of your efforts in Dumfries, efforts for which you are due that town's most heartfelt thanks, and your fellow practitioners' congratulation . . ."

He paused again. Sutherland bowed slightly towards him. "Can you tell us, sir, how you conclude that the diminution of cholera was caused specifically by visitation, and not by the clearance of nuisances which you said you also instituted, or even by the epidemic having run its course in that place?"

Gadd was all politeness. Leggate listened, eager to see how the inspector would withstand the same line of attack that Gadd had made on him when he had first spoken to the Medical Society. Sutherland faced Gadd, who remained on his feet, tall, now tilting his head back a little to regard the inspector through the spectacles that perched halfway down his nose.

"I believe, sir, that every one of the measures I described is necessary. Everything that contributes to abating filth and other conditions which predispose towards the disease were of benefit in Dumfries, as they would be here. I laid especial emphasis on the experience of visitation in Dumfries because I have observed how little has yet been accomplished here by way of clearing away muck garths, diminishing the numbers of pigsties, preventing people from piling excrement under their windows, limewashing houses, or providing places of refuge for the sick, since I was in your town in October last. I described medical visitation in detail because here, with the neglect of so much that should have been done earlier, it is the swiftest remedy available,

one that would allow the townspeople of Middlethorpe to regain some confidence in medical men in this town."

People held their breath. Nobody had heard Gadd spoken to in this way.
All waited to see how he would receive the reprimand.

"I thank you for this clarification," said Gadd.

A pause. Would he retaliate?

"Now, gentlemen," he said, "are there other questions for our colleague
from London?"

Leggate's mind was in motion. Why ever didn't Gadd point out the error
in logic? he thought. The man shows us a diminution of cases of cholera in
Dumfries, and points out a mere correspondence with putting up a dozen
men to make calls and give so-called treatment. If anything should be clear
about this disease, it is that no treatment has been effective. Is this cause?
Leggate scratched his forehead.

Treatment as compared to no treatment, he thought. Why doesn't Sutherland give the comparison?

Leggate opened his notebook. While others asked questions he scribbled
rapidly, formulating the question. He crossed out several versions, until he
had something adequate. He waved his book to catch Gadd's eye.

Gadd said, "Dr Leggate."

Leggate rose. "I thank you, sir, for your interesting evidence of the hundreds of cases of diarrhoea discovered in Dumfries, and how, once discovered and treated, these seldom passed into cholera. I wonder, sir, whether
you were able to tell us how many cases of diarrhoea that were not treated
passed into cholera?"

Leggate paused. It was as if he heard the echo of his words "not treated"
in the room. "Let me put my question another way," he said. "In Dumfries,
were cases of premonitory diarrhoea discovered to whom medical treatment
was not given? And if there were such, could you tell us what proportion of
these untreated cases of diarrhoea passed into cholera? From your remarks,
I take it that among the untreated this must be a large proportion. Would
not this large proportion be the apt point of comparison with the small proportion that passed to cholera with treatment? Is not a comparison needed
to show the efficacy of this treatment?"

Leggate sat down. He sought to catch Gadd's eye. Gadd gave no hint of acknowledgement.

"The figures I gave were cases from house-to-house visitation," said Sutherland. "Its express purpose was to give medical treatment. I must assume, sir, that you are not asking me whether visitors withheld medical treatment, for that would be criminal. I take it, instead, that you must mean to ask whether cases were discovered with full cholera already developed who, when questioned, admitted to premonitory diarrhoea that had gone untreated. I can tell you, sir, that a large majority of them did admit to such symptoms."

Leggate did not reply. The man will only reason backwards, he thought. From cases of cholera back to previous diarrhoea—three-quarters of the population have diarrhoea, it doesn't mean anything. Why won't he reason forwards, from premonitory diarrhoea, and compare the treated with the untreated? Is it ignorance? or pigheadedness?

There were more questions, more confident responses, anxious discussion, outbreaks of unruly noise, occasional bad temper. When Sutherland left, Leggate stayed so that he could talk with Gadd.

Premonitory diarrhoea, he thought, the new abracadabra. Of course he doesn't know that ordinary diarrhoea is from an utterly different species of planticle, which has much milder effects, and grows differently in the intestine. Hundreds of people with symptoms of diarrhoea offered free opium, or dosed with calomel, and behold the cholera evaporates . . . Why didn't Gadd press home his point? I know from Ringers Beck that the cholera takes hold in two days, three at most. Two weeks of premonitory diarrhoea, passing into cholera, what balderdash!

Most of the committee and guardians had also remained behind—people with an urgent need to make themselves heard. As Leggate walked round the room, he heard, from here and there, samples from a babel of opinions.

"There's no cure, so why bother to diagnose it?"

"Depravity, that's what's brought it on us, wipe that out, and the disease is gone . . ."

"Stop any more coming in by ship, that's what I say . . ."

"If he's been successful, we ought at least to do what he recommends."

"Send every case off to an isolation hospital . . ."

"I don't know about limewashing, it's handwashing for me, can't stand to think that anything I touch has been touched by one of them . . ."

After ten minutes Leggate, too impatient to wait longer to talk to Gadd, too irritated to add to any of the knots of eager opinion, left for home.

Leggate was barely able to sleep. Before breakfast he rushed to the Ship Tavern, where Sutherland was found addressing two boiled eggs, pausing from time to time to scribble with a pencil in a large leather-bound notebook.

"Excuse my interruption, my name's Leggate. I was at the meeting last evening. I wanted to ask a question, from your experience as Health Inspector."

"You're the man who asked what proportion of untreated diarrhoea cases in Dumfries passed into cholera, are you not? I have no doubt it was a large proportion. You question cholera patients here. You'll find the majority admit to premonitory diarrhoea. I can't be long, but sit down for a moment if you want me to go over it again. The case is as strong as could be."

"I wanted to ask something else." Leggate sat down. "You said in your address yesterday that you had no doubt that poor diet, like the plums eaten by the sailors on the ship from Hamburg and unwholesome water, predispose to the disease. I wanted to ask: Have specific instances of the effects of water, I mean, come to your notice? I have been doing research on water. I'm a member of the Sanitary Committee here."

"Filthy water certainly contributes, no doubt of it. Drains, sewers, dampness, there's no doubt."

"But specific instances?"

"Dumfries itself is an instance; the place in the Scottish lowlands I spoke of last night. I asked where people habitually drew their water. You would scarcely believe it. From the river, just below the outlets to the common sewers, there they were, drawing water—a most injurious and unnecessary practice. I used my influence to put a stop to it."

"Could your stopping it have caused the cessation of the disease that you described, between the 10th and 18th of December?"

Sutherland had spoken carelessly, as to a schoolboy. Now Leggate saw his eyes narrowed, regarding him: "Its effects combined with the other measures," said Sutherland. "We must neglect nothing. Now if you'll excuse me, I have to meet again with the guardians, to draw up regulations."

In the weeks after Sutherland left the town, some few more people came down with cholera. Leggate saw none at first. He started his experiment. Every day each lodging house that had agreed to take spring water was visited by a water cart. Every second day Leggate visited his houses, both those that took spring water and those that did not.

There were more cases than the previous autumn, but still they were infrequent. Opinion in the Sanitary Committee grew that, despite the new regulations Sutherland had laid out, existing measures were, on the whole, sufficient—nothing new in the way of dispensaries, or hosing down streets with fire engines, or limewashing of houses, or appointing of medical visitors was necessary. Such measures were, they agreed, a kind of fussing. To be sure, more could always be done, but advances had been made since 1832. The new procedures that Sutherland had laid out were expensive. Edwin Chadwick, who was the moving force in the Board of Health, was also behind the Poor Laws, which constrained parish guardians throughout the land to use every economy. The Middlethorpe guardians appointed two extra men to work on clearing nuisances, and decided to hope for the best.

Chapter Fifty-five

At the very end of August, when Leggate was beginning to think that the summer visitation of the disease would be no more severe than that of the previous year, it happened. It happened on a single day—people began falling ill with every terrifying sign of cholera. As if to punish the town, the epidemic struck with unheard-of violence. Not one case here or there, but hundreds. Like a tidal flood, intense fear rolled through the town. In the first week of the disease's exacerbation, the registrar recorded 398 deaths.

There was no time to set up visitation. People found for themselves a different solution: to escape. Queues were everywhere, in the railway station, on the passenger quays of the shipping lines. Carriages, carts, conveyances of every description were on the roads out of the town, slowly overtaking the foot procession.

"People are leaving," said Marian. "Caroline has taken her whole family."

In the greatest apprehension to breathe no longer the vitiated air of Middlethorpe, Caroline had persuaded her husband, as well as her mother and father and their family, to leave for Derbyshire, where a relative had a house.

"You should leave too," said Leggate. "I can't have you and Frederic exposed."

"Leave, for where?" Marian said.

"I've discussed it with your mother. She thought you could all stay with William, at Hadby Grange, and that would probably be all right, but I convinced her that, with Frederic, Scarborough would be better, at her brother's house there. So I thought you should go with your mother, and your aunt too."

"You mean you'll stay?" she said.

"Of course I must. I understand how it works, and I'll be careful."

"If you understand it so well, why can't we stay? We'll only take the spring water," she said.

Leggate saw his wife peering at him, waiting for his reply.

"I see," she said, when he did not answer. "You're banishing me. You want us out of the way."

"Scarborough would be best."

Now it was Marian who did not reply.

"When you get there, find out where the water comes from," he said, "and drink only the spa water, or from a deep well, and make sure there are no drains nearby."

"Is this one of those junctures," asked Marian, "at which a wife must obey?"

"It is," he said.

Marian was silent. He saw that now she would not meet his eye, but looked at the floor as if examining its pattern.

Leggate saw that more was needed. "This will be the last push," he said, hoping that if he were apologetic enough, this might soften her indignation.

"You promise," she said, "that afterwards you'll write it all up straight away. Or if you can't, you'll dictate to me and I'll write it. Otherwise I refuse to go, or else I take you to Scarborough with me."

"I promise."

Leggate saw Marian and Frederic, Mrs Brooks and Miss Mitchell, nurse and three servants onto the steamer for Scarborough. Standing on the quayside, looking up at his wife at the ship's rail, he tried to swallow the knot of loss in his throat, to absorb the possibility that this might be his last sight of her, and this her last sight of him, that he might have made a mistake, so that the disease might ambush him, that he would not survive this epidemic.

He saw sadness in her face, and determination, as she tried to smile at him. As the warps were slipped, he waved to her. Then with gentle thrashing of paddles the boat made sternway, eased into the river, turned its stern upriver, then slowed.

As the steamer faced downstream, and first one then both paddles started more purposefully to churn, quarter ahead, she waved, he waved, and the boat headed towards the great estuary. As he lost sight of her, he walked to the end of the quay and continued looking south as the boat, billowing a cloud of smoke over the town, grew smaller, then, at last, turned eastwards towards the sea, and was out of sight.

His mind turned to the task ahead. The pang of loss became relief that she and Frederic would at least be safe, and—a thought that he did not like but could not inhibit—should I have married? I need complete devotion to this. Is not this tug of wife and child a distraction? What about other scientists? Magendie did most of his work before he married. Science isn't for everyone, but it demands utter devotion, the thing's almost impossibly difficult as it is, even without the cradle in the house that draws us away.

His marriage had not been altogether the sustenance he had longed for. True, he felt he could dedicate himself and the fruits of his work to her, but what he had not anticipated was the need she seemed to provoke in him to justify himself. Now she was off to Scarborough, he would not need at least for a while to worry about that, not need to think about whether she wanted him to be spending time with her, not need to offer explanations, not need to propitiate.

At the times when it comes right, she can be so... everything, he thought, mentally acknowledging his wife's making of their home, her tenderness, her warmth. But she also makes ... demands! He shocked himself at the formulation of this idea.

So severe were those first days of the epidemic, so alarmed was the town when with each new day the count of deaths continued as high or higher than on the day before, that the mayor called a meeting: the Town Council, the guardians, the Sanitary Committee—all together in the council chamber.

"It's as grave as can be," he said. "In a week one in every hundred of the town has passed on. If it lasts for ten weeks it will be one in ten of us."

Dr Withersall expressed the opinion that in proportion to the outbreak being severe, it would be short-lived. There was some assent to this proposition, but one alderman stood up and said, "Two medical men, one a member of the Sanitary Committee, have already died, not to mention a member of the town council, and a banker. While the doctors tiptoe about, it's clear as daylight it comes from the dock area, brought in from the Continent. We should cordon off the area. The rest of the town's cheek by jowl with sailors, prostitutes, and I don't know what, putting all of us at risk with their filthy diseases."

The mayor cut the recitation short. "We're on the critical list, and no mistake," he said. "I've telegraphed to London, to the Board of Health, to get that Dr Sutherland here again. No—I won't hear objections. It's already done, bitter medicine, but it's my bounden duty. He's got more experience with this business than anyone. We must leave no stone unturned."

In the hurry of his many calls, Leggate wanted to be everywhere. He called to see Amy, to reassure himself that the members of her household were complying with instructions.

"Amy's with Connie, to lay out Connie's aunt. I am looking after the baby," said the housekeeper.

"Laying out? I didn't know they'd be laying out! Do you know where they are? I must go immediately."

The housekeeper did not know, but Liza did. She would go with him. In an instant they were running through deserted streets. They found the house, knocked. Connie opened.

"Connie, Amy—I came to warn you, have you finished? Finished laying out the body? Have you eaten anything?"

An older woman appeared behind Connie, and stood wiping her hands on a none-too-clean apron.

Connie replied: "We're just finishing, then we'll have a bite to eat."

"That's what I came to say," said Leggate. "Don't eat, or if you do, wash your hands first, very carefully, with the spring water, before you eat, and take off your aprons first."

The older woman said, "Whatever is this?"

"This is Dr Leggate," said Connie. "He's been telling us how to avoid the cholera."

"Might as well tell us how to avoid being born."

Leggate was inside, explaining: "If you come in contact with a victim, don't eat until you wash your hands. Blankets, clothes . . . all get soaked. The evacuations contain tiny seeds that you can't see. They get on hands, and then if you touch anything, like food, you take them into your mouth, they grow inside you and that's what causes the disease."

Dutifully, Liza went back to the house in Friars Walk, carried back a jug of spring water. Leggate told them that linen they had used should be put in a bucket to boil. The food that had been prepared was thrown out. The women washed their hands. Leggate left behind him an awkward atmosphere, having intruded into this womanly event. He had compounded the sadness of bereavement with the terror of vulnerability.

From the beginning of September, Leggate slept only four hours a night. Sending Marian and Frederic off to Scarborough allowed him time. He was out all day, burned his candles late into the night to keep his meticulous notes. The opportunity was not to be squandered. He gave priority to the lodging houses of his experiment, those who every day received the spring water— also to those who did not. His first cases were of two women in a house that had not been taking the spring water who had collapsed with unmistakable signs of cholera. Leggate showed an attending woman how to make his hydration fluid from spring water, to each pint add a teaspoon of salt and a tablespoon of sugar.

"And where should we get sugar?"

"I'll bring some. The patient must drink as much as possible. If vomiting, better to take small quantities often, rather than a large amount infrequently."

Then a rush of cases broke out in the houses that did not take the spring water, but so busy was Leggate that for none of them could he steal time to attend carefully to supervise his new hydration therapy. Nor had he arranged with any other doctor to record cases with which he could compare those to whom he gave hydration.

He was able only to instruct those who attended the victims, to show them how to make up the hydration fluid. For some it was already too late. On the attendants of others he impressed that they must get their patients, by any means, to drink as much as possible of the fluid.

"If the patient will drink tea, or gruel, or anything, that is well, so long as the fluid is from the spring water, but he must drink. And cleanliness; when attending a sick person you must wash your hands before eating."

On the 13th of September, Dr Sutherland came back to the town and made an inspection. He met in emergency session with the mayor, the guardians, and the Sanitary Committee.

Reports were given. The registrar announced 905 dead in the first two weeks of the epidemic. The chairman of the guardians reported to Dr Sutherland that under the agreement made at his visit almost two months previously, they had taken all legal means to remove certain nuisances that had been identified, but the magistrates had been indisposed to convict offenders. Other reports were made. Then one of the guardians, a pious man, rose to say that Divine judgement had fallen, and that they were utterly beyond human aid.

When Dr Sutherland rose to speak, he reminded the meeting of the agreements they had made.

"If I were to represent the situation plainly," he said, "I should have to say that not one step that was agreed has been taken since I was last here. Nuisances remain, no houses of refuge have been set up, few dispensaries are open, medical attendance is insufficient, no house-to-house visitation has been organized, no cleansing staff has been employed, no houses have been limewashed."

Dr Sutherland reported his findings by telegraphic dispatch to the Board of Health. The Board in consequence issued special regulations: The town was divided into twelve districts. Two medical superintendents were appointed, as well as fifteen medical officers, and fifteen other medical visitors. Ten day-dispensaries and three night-dispensaries were set up, as well as two houses of refuge. An Inspector of Nuisances was appointed, together with a cleansing staff of twelve men for limewashing houses.

The desperation felt generally, from the mayor downwards, as well as the coercive force of the new regulations prompted the guardians and Sanitary Committee at last to act. Several doctors had died, and others had left the town, so salaries were agreed to bring in medical men from Edinburgh. A partial visitation was begun on the 17th of September, with hopes that full visitation would be in place within a further four or five days.

The start of visitation coincided with the start of a spell of hot cloudless weather. A torrid sun blazed down. Fumes from whale-blubber boiling no longer wafted through the town; instead, the smell of burning tar hung everywhere. Dr Wemyss, a member of the Sanitary Committee, had set up tar-burning to purify the atmosphere. Its most noticeable effect was to replace the hot oppressive air with hot acrid fumes.

Ten thousand people had left Middlethorpe. Hammering from the ship-yards had ceased. Shops were closed. The streets attained a deathly quiet—the only sound a melancholy clop of hooves, and the clang of the passing bell as another makeshift hearse went on its way to one of the common pits that had been dug to receive those many dead who had neither friends nor relatives.

Leggate had been appointed medical visitor in a district near the docks where he was already working. The lodging houses of his experiment could be included in his visitation. In the next four weeks he visited every house in his district.

With his experiment on the lodging houses, and now with the official vis-itation, and with such of his own patients who had not left the town, his sleep fell to two or three hours a night. He had planned to give at least twenty patients his hydration treatment, to stay close to observe minutely its effects. But such was the intensity of the epidemic that he was unable to make close observations on even a single one of these. So although the preventative exper-iment with the water carts would be properly completed, the therapeutic study of rehydration would fall short, another casualty of the disease.

Chapter Fifty-six

On the first day on which Leggate had taken part in the official house-to-house visitations, the 17th of September, he had gone out at six in the morning. He had his houses to attend, three people were cases of cholera. One, whom he had left at two o'clock that same morning, was responding only slowly to hydration, he vomited frequently. Leggate feared the gains of fluid were losing the race against the losses.

On the second day of visitation Leggate resolved that, even though he had attended all those in his houses who had fallen sick, he would question every single person once more. He felt the responsibility of his appointment as Medical Visitor.

I'll follow visitation instructions to the letter, said Leggate to himself, even though they're ridiculous, follow them because they are ridiculous and show them to be so. Unnecessary nonsense, there'll be great manufacture of symptoms with free opium going round. But the law's laid down, I'll do it anyway.

So he enquired about all looseness of the bowels. He offered medicine for each such complaint. He was pleased that some refused it; he noted these cases carefully too. They would allow him to calculate just how effective was treatment in preventing diarrhoea from "passing into cholera."

What an absurd expression, he thought.

By the 21st of September visitation was fully in place. Ten days after the start of visitation, the disease continued to claim its victims, though perhaps at a lesser rate. But Saturday, 28th of September, saw no respite for Leggate. He had two new cases with full cholera symptoms, attendants to be instructed, people to be moved to other rooms. By three o'clock he had attended the better part of a hundred people. He wrote prescriptions which the recipients could take to a dispensary, for something free: twenty grains of opiate confection mixed with two tablespoonfuls of peppermint water, to be repeated every three or four hours.

He next saw those of his regular patients to whom he had the previous day promised a visit. He saw, too, those not fully recovered in his other houses. He had planned to make official visits to several houses that were not in his experiment, but was prevented by an hour with a patient who had come down with cholera in one of his houses that had been taking the town water.

It was seven o'clock in the evening when he returned home.

"Give me the messages from patients, if you would," he asked his maid, Elsie. He had shown her how to write them, with the time, the complaint, the address. Elsie gave them. There were three.

"Waiting for hours, some people," he thought. "It won't do."

"And sir, Mr Halliwell's maid came. Mr Halliwell's been taken ill, they fear it's the cholera. Though Dr Newcombe is his doctor, sir, he wanted to see you."

"Newcombe's away. In London, at a funeral. His brother, with the cholera. Did they find another doctor?" said Leggate, in alarm.

"I don't know, sir. She was here four or five times, asking for you," said Elsie.

"Four or five times! I'm at the Halliwells'. Ask the groom to go round to these other people, say there have been many sick today. I will try to come tonight, or if not, then first thing tomorrow."

When he reached the Halliwells', Margaret was pale, distraught.

"It's the cholera," she said. "Collapsed, blue skin, vomiting and the other . . . just as people say."

"Have you had a doctor to him?"

"I got Dr Prentice at last. He left medicine to give every hour."

"Let me see him, and show me the medicine."

"Thomas kept telling me that I had to give him lots to drink, and I did, but he kept bringing it up immediately. I didn't know what to do. Then, he had terrible, terrible cramps. But since the doctor came he's been quieter."

Dosed to the eyeballs with opium, thought Leggate.

They went into the bedroom. Halliwell was insensible. His bluish dark-ened face, once so familiar, was now grotesque, unrecognizable. He has reached the stage of collapse. He was scarcely breathing. No pulse could be felt.

"Halliwell, try to wake up." He shook him by the shoulders. "You Must Wake Up," said Leggate.

"Make me up a drink to give him," said Leggate. "A pint of the spring water, a level teaspoon of salt, and a level tablespoon of sugar. Quick as you can."

Leggate tried to rouse his friend, holding his head in his hands. Halliwell's eyes opened, a tiny fraction, and then closed again. "Le . . ." The sound was inaudible.

Margaret brought the drink. Halliwell lay on the bed, immobile.

"Help me get him to drink something," he said to Margaret. They raised him to sitting position, and tried to pour fluid between his lips. "You must drink this," said Leggate, "if you can."

They tried again, and again, to hold the cup so that Halliwell's lips would open. The fluid ran down his chin, onto his chest. A little went in. He coughed feebly.

"Wake up, you've got to drink, remember."

"I did wrong to let him sleep?" said Margaret.

"No, no, you didn't do wrong." They lowered his shoulders back onto the bed. "Is this the medicine?" Leggate took up the bottle on a table beside the bed, opened it, sniffed, touched a drop to his tongue.

I was right, he thought. The disease compounded with opium. No wonder he's insensible.

"Here, help me again," he said to Margaret. "We must try to get him to drink."

Leggate had more fluid made up—they should try to get large amounts in. They tried again, and again managed to get some fluid down.

Leggate did not hint to Margaret that her husband had been mismanaged. But his actions spoke more loudly than anything he might say.

"We should have been giving him this fluid to drink all this time, instead of the medicine?" said Margaret.

"Sometimes the medicine helps."

The two tried again and again—for an hour. Halliwell suffered a further purging. They changed the bedclothes. He could barely swallow, but bit by bit, some fluid went in.

Then Leggate thought of Latta's method of introducing saline into a vein.

I've no suitable equipment, he thought, don't have it at home either. Could run back home to find a lancet to section a vein—no, take too long, Halliwell must have a razor, could probably use that, then find a tube to insert into the vein. What to use as a tube? The saline couldn't just be poured into a funnel, not enough pressure, it would have to be forced in, with some sort of pump—like the heart. If only I'd thought of this in advance, could have arranged the equipment that Latta used, or O'Shaughnessy. What did they use? Some kind of patent enema syringe, I read about it when I thought of practising the method, and a silver blowpipe pushed tightly onto the syringe nozzle, or something like that, to insert into the vein? Or we had syringes in Paris, that Magendie and I used to inject mercury for our experiments. I was going to get all that kind of stuff, for my new experiments on rabbits. But when the experiments became irrelevant, I didn't do it.

Pump into a vein, he thought, and then . . . Eureka! What about one of Halliwell's model steam engines? the one I saw that he gave to Sophie had a little tube, so that if you turned the flywheel it would also pump. Maybe that would do it.

"I want to try something," he said to Margaret. "It may not work, but I need to see if we can't get some of this fluid directly into a vein."

Margaret looked at him, unbelieving. "Of course, do whatever you can."

"You keep trying to get him to drink, get one of the servants to help you. I need to go into his workshop to find a tube to insert into a vein. I'll ask the cook to make up five or six pints more of this fluid, and boil it, and then stand the jug in cold water to start cooling it, because it's got to be the same temperature as the body . . ."

Leggate rushed through the house, found a razor, gave his instructions to the cook, asked the nurse to find the toy steam engine that Sophie had been given, took two lanterns to Halliwell's workshop to look for tiny tubes that might be suitable.

As well as the little engine Halliwell had given to Sophie, Leggate found four more. They might have worked. They had outgoing tubes that could be inserted into a vein, but Halliwell had not built one with a tube at the other end that would draw in water. Like the one he had given to Sophie, they were primarily to show how the steam-engine action worked, by blowing into the tube, to see the flywheel go round. When worked as a pump, they just sucked in air from around them, and could blow out bubbles from their tubes under water, as Halliwell had demonstrated. Leggate looked again with a lantern through the workshop, trying to find something suitable.

After a frenzied half-hour he gave up. Halliwell had made the little tubes specially for each engine, there weren't any ready-made. Nor was there anything like a syringe to hold the fluid, or even a bag that could be squeezed. Might Newcombe have an enema syringe, he thought. And perhaps a silver blowpipe? Send round for it ... I forgot, he's in London, and his housekeeper lives out, could I break in? ransack his house for them? But why should he have a blowpipe? It must fit perfectly, if I introduced air into a vein that would certainly kill him. Need to prepare all this kind of thing in advance, no good trying to improvise with the wrong things.

Every doctor in the land equipped to let blood out of the veins, he thought, although it's never needed, but hardly a one of us equipped to put fluid in, although now nothing could be more important.

Better see how he is; getting fluid into his mouth probably still the best option, some was going in, and he's no longer vomiting, but what we don't know is whether it goes in faster than it comes out.

He returned to Halliwell's bedroom.

"I'm sorry," he said. "There was a chance it might have worked, but it won't. I don't have the right equipment. I think it's best if we just keep getting him to swallow."

"He's had another ... purge," Margaret said, looking downwards, ashamed. "We've cleaned him, and remade the bed."

The housekeeper, who had been helping Margaret, left, and she and Leggate kept trying for another hour to force fluid between Halliwell's lips, thinking they succeeded in getting him to swallow small amounts, until at last, not even the slightest indication of swallowing occurred.

"Do they ever recover from this stage?" asked Margaret.

"Some do recover," said Leggate.

But the vital forces of the body had ebbed, the soul was ready to depart. Leggate knew there was little hope. Just before midnight, Halliwell gave what seemed like two slight convulsive sobs. Soon afterwards he died.

Leggate, fighting to restrain his face from any unbecoming gesture, folded his friend's arms on his chest.

"I should have been here," Leggate said, unable to look at Margaret.

"You didn't know."

"He was the person I was closest to," he said.

"He was all the world to me."

She lay on the bed, nestled her face against the furrowed face of her husband. She let forth a flood of tears.

Through the night, Leggate sat with Margaret.

"When he went down with it, he kept saying that I must give him water to drink. He kept saying 'Leggate says...' He believed your theory. But when he drank at first, he would vomit it all back, almost at once, so violently."

"It's impossible when that happens," said Leggate.

"At first he could use the chamber pot. But then I had to keep changing the bedclothes." Margaret looked downwards again. "We washed our hands, you know, the maid and I, afterwards, just as you said when you were explaining your theory."

"I know."

"Then when that doctor came and took charge, said the medicine would be the ticket, and Thomas was quieter, went off to sleep, it seemed a relief— I shouldn't have believed him. Should have kept him awake, kept giving him water."

"Sometimes it makes no difference. The disease gets such a start, so much fluid is lost, and can't be replaced before the person is too far gone. That's what must have happened."

"When the purging started at first, he kept saying, 'It must have been that cordial.' You know how careless he is, just trusted everyone. He was at the harbour office on Thursday morning, about some lock gates or something. I said, 'You don't need to be doing that. All that kind of work is stopped.' But he went anyway—you know what he's like."

She became silent as Leggate realized she had heard herself use the present tense, the tense that for him could be used no longer.

"Well, you know how hot it's been," she continued. "And the others got in beer at midday. Thomas doesn't drink, you know that, but the harbour master's wife had given her husband some homemade drink of some sort. The harbour master offered it to Thomas instead of the beer. He said he took it without thinking. He kept saying 'Why did I take it?' and 'Better to have drunk beer,' and 'It didn't taste too bad. They called it cordial, though the harbour master said there was no alcohol in it,' and 'The others laughed and said they'd never found a use for it before.'"

"So that was it."

"I can't believe he's gone. Won't he come to? He was with me yesterday."

At dawn, Leggate returned to his house. He went to his bedroom. Having done all he could to comfort Margaret, he restrained himself no longer. He lay face down on the bed, and sobbed.

"Of all men, why him?" he thought. "The best, the kindest."

Leggate could not sleep. If I'd thought in advance, he said to himself, I should have tried out Latta's method and been ready, got one of those patent syringes and the other stuff. Too much else to do. Or I could've got Halliwell to make me a suitable pump, with one tube to be inserted into a vein, and one to draw up the fluid . . .

A joke of Halliwell's came to mind: "If disease comes with trade," Halliwell had said, "perhaps God is telling us something about trade." But that's not it at all, thought Leggate. Disease comes with befouling ourselves with our own waste that we think will do no harm. We're each part of the larger world. That's what God is telling us.

He rose from his bed. He paced. He went to his study and tried lying on his couch.

If only I'd come back at three in the afternoon, as I'd intended, he kept telling himself. Should have come in to see my messages. I've caused this, and Margaret . . . and his children practically orphans.

In the sadness of his own failure, and of loss, he wept.

Chapter Fifty-seven

After seven weeks of the epidemic, Leggate turned to compiling his tables of figures. In the week after Halliwell died the numbers of deaths from cholera started to decline, and by the week ending 18th October, the registrar reported that the number of deaths in the town was down to ten, in the next week none. The disease had exhausted itself.

Sutherland was notified. He replied that in two weeks he would come again to supervise the collection of figures, so that his report could be made.

Leggate felt a great weariness, from the death of Halliwell, from fatigue, from the sight of so much destruction of life with so little done for most of the cases. But he was determined to make sure all the results from his experiment were clear and collated.

The Tafts Spring water had protected those in the lodging houses who had taken it throughout. In them only one person had gone down with cholera. Leggate heard about him soon after he collapsed: Albert Tonks, a casual labourer at the docks, who spent his meagre wages in the tavern. On questioning, Leggate found that he'd been unable to get beer and had drunk town water from a standpipe by the dockside.

"Where else is there?" Tonks had asked.

"You remember I'd instructed you all only to drink the spring water delivered to the house, and to take a bottle with you if you were out for the day."

"You mean that did it? The water?"

"Now what you have to do is to drink as much of this fluid as you can, even if you vomit it up, keep drinking it, and you'll have a good chance."

"Ain't you going to give me medicine, the opium medicine?"

"No, I'm not. Drinking the fluid will help the cramps, and even if you're in pain, it's better to stay alert."

Leggate had instructed a woman how to care for him, saw to the handling of soiled bedclothes, moved those who shared the room to other rooms. He had made up bottles of the hydration fluid, told the woman to make the man drink it every quarter-hour, whether he wished it or no. He had given the woman soap, told her she must wash her hands before eating or drinking anything herself. The man had survived. No one in attendance had been affected.

He wrote figures representing the number of people in each house who had remained living there throughout the epidemic. Under these figures, he wrote others: those in each house who had suffered from diarrhoea, those who had clear symptoms of cholera, and those who had died. Then he wrote his conclusions, the gift his soul willed to the world.

PREVENTION

In the houses which took only spring water, those who were instructed and complied with the instructions, of 207 persons, 16 had symptoms of diarrhoea, none had symptoms of cholera, and none died—o cholera cases out of 207. In one of these houses one man (Tonks) had symptoms of cholera but was discovered not to have complied with instructions. A number of people left the town, and so reduced the original number in the experiment.

In the houses which did not take the spring water, of 232 persons, 126 had symptoms of diarrhoea, 19 had clear symptoms of cholera, and 7 died of it—19 cholera cases out of 232, or 8 per cent.

THERAPEUTICS

In the houses of my experiment, 15 who had clear symptoms of cholera could be given proper nursing and my hydration solution. Of those, 4 died, 27 per cent—better than the general rate of about 50 per cent who die once they have taken the disease, though I shall have to get the figures for our town overall, when they can be gathered. So if losses of fluids can be replaced, the body's own curative powers somehow manage to throw off the planticles. One should use Latta's method for those *in extremis*. If the blood's fluids can be kept up by any means, leakage ceases after a few days and the patient recovers.

DEATHS

There were 4 who died with treatment, but 4 others died to whom I was too late to give hydration, so 8 people died of the 232 in my houses not taking spring water, 3 per cent—not much more than half the proportion dying in the whole town.

Those who took the cholera in my houses would have been far more numerous, I believe, in their cramped and disgusting conditions had I not cleared the rooms of those who went down, and specified how the sick should be attended, and how attendants should wash their hands. This fact works against my figures, but even so the case seems conclusive.

I shall add the therapy figures from my houses to those from my practice, he thought. And I must compile final visitation figures for the Board, and for Sutherland, they'll want to know those. Administering opium for diarrhoea indeed! What utter nonsense. I shall tell Sutherland so when he comes, although perhaps not, better await my publication.

And Marian, will she not be proud at last?—she didn't want to go away, but it was better she did, and Halliwell . . .

Despite the deaths, despite exhaustion, Leggate now shocked himself by the exuberance of his mood.

I've shown it, he thought. Done what I've striven for, done what I've believed in for seventeen years.

By the beginning of October the weather had returned to its normal patterns following the unseasonable heat, and now at the end of the autumn the usual pattern of rains from the south-west alternating with calmer periods had reasserted itself.

Yesterday a storm had blown out towards the North Sea. Now the wind had veered, to fall light. Leggate walked beside the river, which daily brought the life blood of trade to the town, the river that for seven weeks had carried death to be pumped up 160 feet into the water tower, then to flow through pipes silently throughout the town.

They must all be gone, he thought, the sporules in the river, brought in from India. All back to normal, just our own species. Is winter coming? Or the sporule season ended? Sutherland would attribute it to visitation! Must get the real truth out, the next problem to work out how these sporules live, and how they can be eradicated.

He walked out onto the squat square pier at the mouth of the river, where the pilot boats moored. Alone, he looked seawards down the estuary, linked his fingers together, stretched his arms over his head, then separated and stretched his fingers up as far as they would go. He let out a large breath.

A ship was coming up the estuary on the last of the flood, a column of smoke drifted sideways. The outlines of the ship were picked out in that exquisite detail that only occurs in north-westerly airs when a storm has just passed.

It's done, he thought, tears moistening his eyes. If only Halliwell . . . He would have been pleased to know.

The tears in Leggate's eyes made a blur of the vessel coming up towards him.

Two days later, Leggate took the little steamer north to Scarborough to bring his wife and son back to Middlethorpe. He felt like a soldier returning to civilian life from a cruel campaign, himself marked irrecoverably, having seen things, having done things, that had no part in the ordinary world where ordinary life goes on.

He rang the bell of the house in a quiet street in which Marian was staying, was greeted by a maid who apologized that the master and mistress of the house were both out, but showed him into an airy room, where presently Marian and her mother appeared. He kissed them both, with a sense of meaningfulness that he did not believe they could possibly understand.

"You're all well?"

"Perfectly," Mrs Brooks replied, "the epidemic never touched us here."

"And baby Frederic?" he said. "He's in good spirits, with this healthful air?"

"Thriving," said Marian. "He's asleep just now . . . It's all over then . . . in the town?"

"I believe so," said Leggate. "It's passed."

Miss Mitchell appeared in the room. Leggate went to kiss her cheek.

"You say it's over?" she said.

"I shall be glad to be back," said Mrs Brooks. "Scarborough is all very fine, but I like home."

A silence. "You are looking drawn," Mrs Brooks continued. "Far too thin. You've not been eating."

"My friend Halliwell died," said Leggate, surprised now that with three people staring at him he could not hold back tears. "Sorry," he said, "I'm sorry. I'm tired."

"Oh, John," said Marian, at her husband's side. She squeezed his hand. "Oh, John."

Leggate took the handkerchief she offered, and sniffed.

"He drank some wretched stuff made by the harbour master's wife. I tried to save him, but I was too late."

They were silent.

"I've heard a lot died," said Mrs Brooks, quietly. "Mrs Roscoe, who was a close friend. And Mr Timmins, and that curate of St Giles, Mr Gibbs."

"More than eighteen hundred died," said Leggate.

Another silence.

"So many," said Miss Mitchell.

"Have you seen William?" asked Mrs Brooks.

"I saw him yesterday. He came into town, and said that although in his village there was a terrible outbreak, everyone in his house and farm were safe. I'd told him beforehand that so long as everyone drank only from his well on the west side, they'd be safe, because there's nothing that could contaminate it, and all the drainage is in the other direction."

"Frederic's been quite himself here, the air is grand," said Mrs Brooks. "And we've walked every day on the esplanade."

Leggate looked at his wife, who met his eyes.

"Did you finish everything? Is it done?" she asked.

"It's done. I've proved the disease is spread by contaminated water, without a scintilla of doubt, no loopholes, no qualifications."

"You've got everything you need to publish now," said Marian, back in Middlethorpe.

"I have," said Leggate.

"You remember what you promised?"

"I'll make sure it's sent off."

"And if you get too tangled up in qualifying clauses, I must be called in directly. If necessary I'll go up on the train to London, and give it to the editor myself."

"I will do it," he said.

"How long do you think it will take?"

"Three weeks. Certainly not more than a month. I must get the tables right, and check my calculations, and make sure of the registrar's figures."

"This day next month, then, the 24th of November. And after that you promise to accept my help. I can write from notes or drafts, or take dictation from you—we'll work till it's finished."

"I promise."

"You're sure I can't do anything now?" Marian asked. "Copy tables or notes? I could draw maps. Whatever you set me to—I shall make no remarks, just be an assistant."

"I'll just work steadily, the loopholes are closed—before it wasn't complete, but I couldn't see how. I think that stopped me then, but without my quite knowing why, a good thing really. But it's clear now. Completely clear."

"You said that you lacked assistants here, when people in London had them."

"I did? . . . just bring me tea sometimes, and let me hear your piano, and let me feel you love me."

V.
Finale

From the Old World to the New

There is a great deal of unmapped country within us which would have to be taken into account in an explanation of our gusts and storms.

George Eliot, Daniel Deronda

Chapter Fifty-eight

In November, two publications struck Leggate with especial force. The first was by Warrinder.

TOWN DOCTOR ATTENDS THE DISSOLUTE
WHILE NEGLECTING OTHER PATIENTS

Dr John Leggate, lately come to Middlethorpe, a physician who argued against improvements of the new town water, behaved in a most questionable manner in the recent epidemic of cholera. *The Examiner* has it on good authority that during the epidemic, Dr Leggate gave close medical attention to women of four notorious houses of pleasure in the town, while at the same time, we have learned on equally good authority, that his own patients, applying for his attendance, were told that they would have to wait. Wait they did, for the doctor was occupied elsewhere in the way described; in more than one instance the waiting had fatal results.

One of the establishments to which Dr Leggate gave his full attention was the house in which had lived the late Marie Watson, the woman murdered by Augustus Biggs, in the case in which this newspaper played a part. Since those events, the other women in the house have continued to ply their trade. Three other houses, which Dr Leggate attended assiduously while the cholera raged through the town, were of the very same kind, and in these, too, the women were all saved from the disease, and attributed their escape to this same doctor. Altogether, we have ascertained that Dr Leggate gave especial aid to 28 such women, and all survived the epidemic unscathed.

Readers may wish to compare this number with the 1800 persons in the town who perished from the disease, and they may wonder why these 28 were reckoned by Dr Leggate to be more deserving of his attention than those who died. Readers may also wish to connect these facts with the state in our town, in which from want of attendance by the town's own medical practitioners, as required by the Inspector from the General Board of Health, doctors had to be brought in from Edinburgh.

Whether Dr Leggate was acting according to some theoretic idea that the depraved were more deserving of medical attendance than the respectworthy, or whether he acted from some other motive, we cannot say. What is beyond dispute, however, is that to the lamentable history of the epidemic of cholera that lately visited our town, this episode contributes a distressing footnote.

Leggate read with disbelief.

"In some superficial way correct," he said out loud to himself.

Twenty-eight women in the four houses got spring water in the experiment, he thought. All saved, and they are prostitutes, I suppose, not all of them in the proper sense, but they live off men... other patients kept waiting and one did die, and then turning the knife in the wound, Halliwell died too.

Leggate dropped the newspaper on the floor, gazed at the ceiling.

Doctors were brought in from Edinburgh, he thought, but what an insinuating construction. He must have heard about the spring-water scheme from Millie.

Leggate jumped to his feet, marched down to the kitchen, found the maid, Elsie: "Did a man from the newspaper, Mr Warrinder, come to question you?"

"Mr Warrinder? I don't know that name, sir."

"In the last week or so, a burly man who raises an eyebrow like this." Leggate tried to imitate the expression. "Did he come to ask whether, during the cholera, patients were kept waiting?"

"A man did come, sir. I told him you was quite run off your feet, for weeks. Never seen you so worn out. He said, 'So the doctor's usual patients grumbled?' Of course I said, 'They did not.' Said you went to them as soon as you was back, and I'd give you the messages, so you was going to your patients right into the night."

"And you told him how long they had to wait, and the names of these patients?"

"I beg pardon, but I wouldn't do that."

"Very well, you did right, Elsie." Leggate left the kitchen, morose.

The full blow was to come. When Leggate returned from his last call that afternoon, Marian confronted him before he had removed his boots.

"What is this, John? In the newspaper." She held it towards him. "Twenty-eight women." She did not call them by the name so prominently in her mind.

"Come in here," he said, going into the dining room, and shutting the door behind them.

"You let me believe that all you think about is science."

"You don't think that I have . . . with those women? Surely you don't think that. They are my patients, and part of the experiment. I told you . . ."

"I don't know what to think . . . You have some attraction to them."

Leggate was silent.

"I had to make my theory firm," he said at last, tight-lipped. "I told you about the lodging houses. In science one must be exact. Warrinder is an evil meddler."

There was a long silence.

"I don't want to talk about it," she said. "I'm taking Frederic. I'm going to stay with my mother."

Next morning Leggate heard his wife returning home. She had left Frederic with her mother.

"Are you attracted to women of that kind?" she asked out of earshot of the servants.

"I was thinking of my experiment."

"You aren't telling the truth. And Warrinder, who's taken up with someone in that house, Millie I think her name is, he's found out something from

Millie, he knows something, or he wouldn't be smearing it about the town. Was one of them helping you make arrangements, the one that you like, Annie, is that it? And Warrinder knows of something between you?"

"Amy," said Leggate. "That's her name, and there's nothing between us."

"But she helped."

"She helped arrange with some of the other houses."

"So she's a prostitute and she can help, but I am a wife and I cannot."

Unable to find anything to say to this accusation, Leggate left the room and went to his study. He did not speak to his wife for two hours. At last he went to find her. She was lying fully clothed, prone on the bed.

"I made a mistake," said Leggate. "I shouldn't have done the experiment like that, should have just used ordinary lodging houses . . . but to have you accusing me!"

She did not reply.

"There were twenty-eight prostitutes and nearly two hundred other people who lived in poor lodging houses that took the spring water, some of them of the very lowest kind, and only one caught the disease, and I found he had drunk town water."

She remained silent.

"I saved him with my hydration therapy. And in other houses, out of 232 people who refused spring water, 19 went down with the disease. None with the spring water, compared with 8 per cent who caught the disease without it."

Still no answer.

"Can't you see? I made a mistake. I thought this would be the evidence to scotch all doubters. Even the licentious are safe if they drink pure water—and I was right. But it was a mistake."

"All I can think is of you with a coquettish woman going round the town making arrangements, and flirting—that's what they do, those women, isn't it? Whether they do anything else or not, they flirt. It's hateful."

"Do you have no . . . loyalty? As if I would go with a prostitute; that's what all the gossipers in the town are saying, and now you're thinking it as well."

"I've offered to help, offered a dozen times, but you don't want help from me, you want someone more feminine..."

Leggate left the room. He sat in his study, not knowing how to reply, no longer angry but dejected, humiliated. He felt hopeless, he started to imagine the stories circulating round the town.

I'm ruined, he thought. Even if I bring some action against Warrinder for defamation, or libel, or whatever it's called, he'll bring witnesses to show that what he wrote is true. It would make things worse. I shall leave. Go to France and doctor in some obscure country parish. Or Australia—I could doctor to convicts.

Worst of all, he felt ashamed. As if I've waded through all the muck garths in the town, he thought, then come into the house, smeared excrement on every chair.

Proved, but in a way that can be of no use to anyone, he thought.

The maid knocked and came in. "Dr Newcombe's here, sir."

"Thought I'd better come to see how you are," said Newcombe.

"With Warrinder's article you mean. I made a stupid mistake."

"Malicious man!" he said. "A ray of light, and he thinks of nothing but derision. I shall go and demand a retraction."

"It'll do no good," said Leggate, alarmed now at the thought that Newcombe would know Warrinder had been Marian's suitor. "Sentiment will be on his side. He has a nose for sentiment."

"I'll write a letter, giving the proper facts."

"He won't publish it," said Leggate. "Gadd might do some good. He could speak to Warrinder."

"You haven't heard?" said Newcombe. "After the meetings with Sutherland last week—you must have been there—Sutherland told Gadd that he regarded him personally, along with the guardians and the committee, but especially Gadd, as principally responsible for non-compliance with the Board's regulations before the outbreak, and for the deaths of eighteen hundred people. He's said he'll be reporting this to the Board of Health, and says he'll give testimony to that effect if an official enquiry is held."

"He didn't say anything about that in the meetings."

"Must have wanted everything calm when he was collecting the figures. I'm told he spoke to Gadd just before he left."

"So not just Gadd, the whole committee," said Leggate. "We're all under censure. I failed there too."

"But you were correct," said Newcombe. "The Board and Sutherland were just following their noses: filth equals disease. And I don't know exactly what else Sutherland said, but in any event Gadd has left the town, gone to Kirkwall in the Orkneys. He has some family there. His practice is up for sale."

Chapter Fifty-nine

Leggate sat alone at breakfast next day, after sleeping fitfully on the couch in his study the previous night. He sipped tea, not inclined to eat. Elsie had put on the table the newly arrived issue of *The London Medical Gazette*. Mechanically, he leafed through it, with scarcely any interest.

What's this? he thought.

ON THE PATHOLOGY AND MODE
OF COMMUNICATION OF CHOLERA
BY JOHN SNOW, M.D.

Every man Jack with a medical degree now writes about cholera, he thought. Used to read them all, now I can hardly look at the titles.

Nonetheless he continued, skipping the summary at the head of the article, and going to the first sentence.

> Writers on cholera, however much they may have differed in their view concerning the nature of the disease, have generally considered it to be an affection of the whole body, and consequently due to some cause which acts either on the blood or the nervous system.

Not a bad beginning, he thought. He read on:

> ... The following are the reasons which have led me to entertain the opinion that cholera is, in the first instance at least, a local affection of the mucous membrane of the alimentary canal ...

From these few words Leggate received a violent jolt.

"My theory," he said aloud. "In the very journal where I should have written it."

It was as if his heart leapt out into the room, leaving a hollow void. Here were matter-of-fact words, such as he himself might have used.

Just the start, Leggate thought. Does he have all the rest too?

His eyes skipped back to the summary. There it was, within a dozen lines of single column italic print.

> *Cholera poison is contained in the evacuations and communicates the disease by being swallowed: illustrations of this in the houses of the working classes—in mining districts. Cholera communicates by drinking water: cases illustrating this. . . .*

The whole thing, he thought. Alimentary mechanism, evacuations, drinking water.

He forced himself now to read, word by word, all seven pages of the article. He glanced at the next two articles, both on cholera, one on its propagation by contagion, one on treatment. He went to his study, lay on his couch.

Finished, he said to himself. Quite finished, the same thoughts, but he expresses them perfectly. Such phrasing: "one feature immediately strikes the inquirer—viz. the evidence of its communication by human intercourse . . . it has never travelled faster than the means of human transit, and usually much slower." Compelling inferences at each step. Towns on main roads are affected as the disease has travelled from India, but not villages that lie a little way to one side of the roads. Why? Because, of course, the afflicted travel along the roads and do not go to villages to one side to leave their evacuations there!

Then the miners down the pit, in confined space. If one catches the disease, he must contaminate the others who take their food down with them. Then examples of drinking water in Southwark, in Rotherhithe, in Exeter, even in York. Where several people have died he finds a cesspool, or drains leaking into their well. Where the disease has been severe, he finds sewers emptying into rivers from which people take their water. The whole thing, entire and complete, perfect . . .

He has the local loss of fluid from the blood into the alimentary tract. He even has the thoughts about growth: "the cholera poison must multiply itself by a kind of growth"—how long was I stuck on that point!

Marian came into the dining room. "We can't go on like this," she said. "Not talking for the rest of our lives."

"I don't know what to talk about, or how to talk about it."

"Don't you see? I feel jealous of you and pretty fair-haired flirting women, when you want to talk to them, and to Caroline, and to Newcombe, and Halliwell about your research—to everyone except me."

"I don't talk to Halliwell any more."

"I didn't mean Halliwell—but don't you see?" She took a pace towards him, though she still could not bear to touch him, wanting to pain him by being out of his reach.

"Don't *you* see I made a mistake!" he said angrily. "As it turns out, all for nothing, for completely, absolutely, nothing. See this here." He held up the journal. "A man called Snow has it all in this week's *Medical Gazette.*"

She felt far from him at that moment, but this softened her. "Someone's published first?" she asked.

"He wrote what I could have written two years ago. Before we were married. I should have listened to you, let you write it—I had more then than he has now—we could have said enough."

Marian was silent.

"I didn't want to let you see what I was writing before it was finished. I wanted you to be proud of me, not see defects. Now I am more ashamed than I can say, mortified, I've ruined our whole life here, all for nothing."

"Pride and shame," she said. "Is that what it comes to?"

She saw her husband was unable to look at her.

Feeling now the nudge of sympathy, she said, "My father used to make me feel ashamed."

"He did?" He looked at her, gratitude in his eyes.

"It leaves a mark, more than a mark, it's been... it's hard to get over shame."

"My father too. I've been thinking, this was to overcome all that, to have gained people's respect."

They were silent.

"Nothing I did was ever good enough," he said. "My father could be scornful. I don't know why it comes to mind now, but when I was sixteen, I'd written something, we had a magazine at school, and I'd had something accepted for it, and there it was in print, I had a really good idea—can't even remember what it was now—but there was my article in print, and I gave it to my father to read, and all he said was 'In the third paragraph here, this sentence doesn't have a main verb.'"

She moved towards him, reached towards him, touched her hand to his cheek.

Leggate came next day to a realization—it was he himself who had thrust the dagger, in one action letting her feel betrayed and excluding her from his research, while all the while he had been bewitched by the idea that she would love him because he was a scientist.

For what? he asked himself. For three magic letters, F.R.S., Fellow of the Royal Society! Ludicrous, when I can't even publish an article, when all those others chatter away in the press whether they have anything to say or not.

He paced the floor of his study. I've not been straightforward with her, he thought. I did enjoy Amy's company. The practice ruined, won't even have income enough to care for her and Frederic. And now, every time she goes out of doors, at every visit, in every mind will be the idea that she's the woman whose husband consorts with prostitutes. If I admit that I like Amy, in so many words, that will turn the knife in her wound.

"I am sorry," he said, when they were in bed, that evening. "I don't know how to say it, but you're right, I think I must have had some attraction to Amy, secret to myself. I hadn't realized."

Marian raised herself on her elbow, and looked at him wanly.

"I know," she said. "I went to see her."

"You did?"

"I can see why you would like her. A good-natured young woman, fair-haired, vivacious, pretty."

"But there wasn't anything."

"I know."

"You asked her?"

"When I went, I'd imagined I would ask her directly—a stupid impulse really—but when I saw her, I could see from her manner that there hadn't been anything improper. She was polite to me, straightforward. She admires you. Before I went I felt jealous, but that's not it really . . . it was . . . all those things we spoke of, when I was ill, after we married. I became too dependent on you. I wanted to rely on you, and take part in everything with you, not be excluded."

"I wanted to give you something, my discovery. Now I've nothing to give."

"Perhaps that's best then, not giving me some thing. Just you."

"You would have liked me to be in the Royal Society."

"I would have liked it for you," she said.

They were silent.

"I'll go and see my mother tomorrow," she said. "And explain all this business about brothels."

"I'll come too. There are no patients for me to see. It's the end of patients here."

Chapter Sixty

Ruminating still over the events of the previous few days, Marian heard the doorbell. It was her brother, who had ridden in from Hadby Grange to see her.

"I got a letter from Mother," he said. "What a terrible thing."

"It's all here in the newspaper." Marian searched for Warrinder's offensive article, finally found the paper, and handed it to her brother.

She waited until he had finished reading.

"John was conducting an experiment," she said. "He had a number of people in lodging houses, including some of those women, drink only water supplied from Tafts Spring—you remember he had handbills printed about making spring water available—and he compared those people with others who drank the town water. It was to test his theory that cholera spreads from drinking contaminated water, and the experiment confirmed he was right."

"And you?" said William. "What about you in all this? However did he think this would affect you?"

He banged the newspaper on a clenched fist. Marian saw him bracing himself, as if he meant to go and fetch his horsewhip from the front hall, then search the house for her husband.

"He didn't think. He didn't imagine this construction would be put on it."

William seemed not to know how to reply.

"We've had words," she said. "A terrible quarrel, awful—he should have let me help him, but in the end it's not his fault, or at least the newspaper part's not his fault, it's that spiteful Warrinder."

"You're sure?"

"Of course I'm sure."

She saw her brother contemplating: "I'll go and find Warrinder," he said.

"There's no point."

"I'll demand he print a retraction, else he's out on his ear. He'll learn who owns his fine building. That'll show him."

"You can't," said Marian.

"Only what he deserves. I'm going straight away."

"Of course you can't do that. The damage has been done, it's not the kind that can be undone. Whatever happens, John's always going to be stained by this, and if you make Warrinder print a retraction, or throw him out of his offices, he'll never keep his mouth shut. He never does. He'll let it be known in the town that he was coerced by us, so that compounded with consorting with prostitutes, John will become a blackmailer and a thug—just the thing for a thriving medical practice."

They sat for a long time. Marian eyed her brother thinking through the implications of what she had said. She had already dwelt in ardent vengeance on the scheme of putting pressure on Warrinder to retract and to crawl. It was a scheme that she had only with difficulty managed to relinquish.

"So he was doing some sort of experiment towards his discovery?" William said at last.

"He was."

"Would he like it if I went to chat with him, to give him my sympathy?"

"He might like that. He needs sympathy just now."

Later that day Marian and her husband went to see Mrs Brooks.

Marian said, "We need to talk to you."

Mrs Brooks looked anxiously at her daughter in a way that Marian had no difficulty in interpreting as questioning whether the rift was healed.

"Things are all right," she said, "between John and me I mean."

"Thank goodness," said her mother.

"In *The Medical Gazette* that came out this week, a Dr Snow published a theory that was the same as John's. John was too late," said Marian. "He won't get the credit now. But I wanted to talk about the things Warrinder has said."

"The business with brothels is all nonsense," said Mrs Brooks. "But the water—it makes perfect sense to me . . ."

Mrs Brooks turned to Leggate. "Does it matter a lot that someone's done something similar?"

"It does matter, but that's not the most important thing. We've come to say that I can't see how we could stay here," said Leggate. "My practice is ruined, and I can't have Marian going about with people whispering and sniggering."

"You'll go to London?" she asked—anxious now.

"That might not be so easy," he said. "Dr Sutherland is making a report of the epidemic here to the General Board of Health. Proportionally more people died here than in any other town in the country. Dr Sutherland squarely blames the Sanitary Committee and the guardians."

"But you showed it was due to the waterworks and the sewage," said Mrs Brooks.

"My finding became pretty much for nothing when someone else published first," said Leggate. "You only need to discover something once. But in any case that conclusion's still not accepted, it takes a while for these things to have effect. The Board thinks the disaster here was because the committee and the guardians neglected regulations we were supposed to follow, didn't clear nuisances, didn't set up medical visitation early enough."

A pause. "In England, nothing about the whole epidemic will be known more widely," he said, "than this town, and the terrible way we suffered. Being a member of the Sanitary Committee, I can't see how in London I could set up practice just now."

Leggate and his wife went to see Mrs Brooks again the next day.

"We've thought we might go to Canada," said Leggate. He wasn't sure why his wife had considered it best that he make the announcement.

"Canada?" said Mrs Brooks. "First William, now you," she said, looking at her daughter with the alarm of a cornered animal.

"William didn't go," said Marian.

Mrs Brooks shrugged her shoulders. "But you will," she said. "And as if you think that him going would be the same as you." She took out a cambric handkerchief.

"You mean he was just going for a while?"

"Never mind what I mean," said Mrs Brooks.

Marian told Caroline of the plan to go abroad.

"Canada! It's thousands of miles away," Caroline said.

"You'd miss me?"

Caroline closed her eyelids.

"Don't be so silent," Marian said.

"I'm sorry."

"Three thousand miles away," said Caroline, looking at her friend with pained eyes.

"We'd go in the spring," said Marian. "Before that there's ice on the rivers of America."

"I am sure the houses have great fires. You'll have a horse-drawn sleigh."

Marian spoke to her aunt that evening, of her husband's disappointment, of the probable effects of Warrinder's attack, and of the decision to leave for Canada.

"It would all pass over if you stayed here," said Aunt Mitchell. "It will pass."

"Perhaps not," said Marian. "Remember Anderson and Ayer, the two doctors who pulled each other's noses, and nearly fought a duel in the last epidemic. People still love to remember that."

Her aunt still remembered the incident herself. She sucked in her breath.

"John thinks it would be best," said Marian. "He says even if he stuck it out, he couldn't bear to think of me gossiped about as the woman whose husband went junketing with prostitutes during the epidemic."

"Your friends won't think that, and his patients respect him."

"Respect—the whole thing was to gain respect, so that he could feel he'd given the world something."

"People visit America now for the fun of it," said her aunt. "Harriet Martineau went. I'll come and see you."

Chapter Sixty-one

The Leggates' baby had an easy time of it until he was nearly a year old. But cholera is not the only epidemic disease. In February there was an outbreak of diphtheria in Middlethorpe. Frederic caught it.

"Can't you do something?" Marian said to her husband.

"I'll do everything in human power and knowledge," he said.

He touched his wife's arm. "It's one of these diseases that takes its course," he said. "I'll sit with him, and notice the slightest change. We're in God's hands."

When Caroline visited the Leggates the day after the disease was diagnosed, Leggate emerged from the sickroom for a few minutes, while Marian went to sit with the child.

"How is Frederic?" Caroline asked. "Shall I go in to see him?"

"Better not," Leggate said. "We don't know how this disease spreads— for all I know it may be in the air. I wouldn't wish it on your children. I'd keep them at home as much as possible until the disease dies out."

"Should I wash when I go home now?" asked Caroline, thoroughly alarmed.

"You can if you like, but I think it's best just not to get too close to any of us here, and no touching of anyone with the disease."

"But how is Frederic?"

"His attack is not as severe as some, but one cannot say. Certain kinds of deterioration haven't occurred, but this disease is more often fatal in infants than older children, so one can't be sure. I sit with him, and try to let him feel my love for him."

Leggate would show Marian the child's throat, and point out characteristic signs of the disease. He would hold down Frederic's tongue with the handle of a teaspoon, which he had rinsed in a glass of water. At the back of the child's throat was the greyish growth of the disease.

"It forms a kind of membrane," he said. He looked at Marian.

"The growth of this grey stuff in the throat," he said, "like a mould of the kind that grows on food, which I studied a bit. It's another species of planticle—this one grows in the throats of children."

"We can't root it out?"

"I have been removing it." He showed Marian how he removed the membrane with one of her smaller water-colour brushes.

"As the disease progresses it becomes more attached, and penetrates the tissues. Then I judge that it best be left in place as it causes the throat to bleed if removed, which cannot be good."

The fact that his practice was a fraction of its former size allowed Leggate to stay at Frederic's bedside almost continuously. He made sure the child took sugar water, a little warm milk. Occasionally, he listened with his stethoscope to the small lungs, back and front. Often he inspected the throat, felt the faint pulse. When Frederic became too feverish, he wrapped him in a damp sheet to cool him. For the most part he sat in vigil.

"I feel helpless," Marian told Caroline a week into Frederic's illness.

Caroline did not go inside for fear of infection, but Leggate had said it was safe enough for the two women to talk in the open air and, since it was a warm day, they had taken chairs to sit in the garden at the back of the Leggates' house.

"When I sit with Frederic," said Marian, "it does no good, because I can't be calm. The thought that he might die, and be no more... I can't even think that, it's so dreadful, but then I can't stop thinking of the possibility."

"Thinking of the illness."

"And not able to do the slightest thing. My fear must communicate itself, and that must make it worse."

"There's John."

"His presence is the only thing that helps," she said. "He tells me to sleep, then sits through the night. When I see him with Frederic, I feel safer."

"Because he knows how to sit with the child."

"He has an inner calm," she said, "as if we really are in God's hands."

A week later, Leggate came to Marian as she lay in bed after another night's intermittent sleep.

"I didn't dare say anything until now," he said, "but I believe Frederic's throat is clearing up. His heart is stronger."

"You've saved him," said Marian.

"Not I," said Leggate.

"You're sure he'll be all right?"

"He's been improving steadily for a night and a day. Come and see, and tell me if he doesn't look brighter."

The child's breathing was easier. Marian went to stand over him. Her smile was answered by one from the child.

Marian went that day to the church that used to be her father's, on her own, knelt at a pew near the front. She looked upwards to see light flooding in colours through the glass above the altar. She thanked God for Frederic's recovery, and for Caroline's children who had remained safe.

Her husband, in a more private way, lay on his couch and closed his eyes.

During his ten-day vigil, while his son would sleep fitfully, Leggate had turned his mind on his ambition and on reaching his goal—a kind of hubris it seemed now visited by the nemesis of Warrinder.

In the end, he now thought, isn't this what it means? When I first met Marian, at her father's bedside, she said, "Shouldn't you act on the disease?" Sometimes one knows how to act, then of course we should act—as I now know how to act with cholera, but often not, and often one only thinks one knows. But we can be there. With Halliwell I wasn't there. With Frederic I have been, and this is what is important: to be there, close by. Not necessarily busy, because for humankind, though we can do some things, so few of the forces upon us can be controlled—but we can be present.

He thanked God for his son's deliverance.

Chapter Sixty-two

The Leggates had reserved a compartment for their journey to Liverpool, across the isthmus of England. Marian sat with her back towards the engine. The nurse sat next to her, with Frederic on her lap. Through the carriage window, Marian looked back towards her mother, her brother, her aunt, Caroline, as they stood on the platform. She moved her hand in a small arc.

Marian and Leggate had said their goodbyes, promised letters to assuage the pain of parting, wondered much if this new course had not been too hastily conceived, or if it were not too drastic. Marian's mother and aunt were resigned, her brother matter-of-fact, Caroline still tearful. Leggate had been to London to see his mother, then to see Margaret Halliwell, who was still bereft, and had been also to see Newcombe—even persuaded him to take over the part he had undertaken as physician to Connie's baby Florence and her household. He had visited the household, and had thanked Amy warmly for her help in making his discovery.

Now, the moment of departure and of separation: the train jerked, a quick breathless puffing as the engine's wheels turned but failed to grip the rails, then another jerk before the carriages began to move more smoothly, and began slowly to pull out. Leggate leaned from the window, waved his arm.

Liverpool was more than they expected—a city made rich by trade in slaves, now multiplying its riches by trade in emigrants, from England, from Ireland, from Germany, leaving the Old World under more or less coercion for the New. From the station in Liverpool, the Leggate family took a cab, which set out into the street to be caught in a mêlée: carts, carriages, horses. They arrived at last at a hotel.

Two days later they were aboard ship. Neither Marian nor the baby suffered from seasickness. Leggate and the nurse were not such good sailors. For three days after their departure, for another three during a spell of rough weather, Leggate felt so nauseated that he wished only to die. Though relieved that an official enquiry by the Board of Health had not been held, nevertheless, for him the journey was to exile, his seasickness an additional punishment.

New York caught their attention. "Paris with a dash of Hamburg," as *The Times* put it that year. There were hundreds of berths for hundreds of ships. Warehouses, the grand houses of the rich, churches, establishments for dining, streets crossing each other at right angles, and shops, and shops. Seemingly in New York one could buy anything. And people everywhere: so many feet, so many hats, so many bags.

The Leggates' inland voyage took them up the Hudson to Albany, then by the Erie Canal to Buffalo, then the Welland Canal, at last to Newborough, principal city in Upper Canada, on the shore of Lake Ontario.

Marian wrote to her mother:

We have taken a house on Queen Street. It is not bad, with a stable in a mews behind, and room for a carriage when we get one. Uncle James is quite established here; it is a comfort that he is near, though he is a few miles north of the town. Little Frederic has not been in the least perturbed by the move. He liked the ships and the hotels—being all together with his parents and the nurse on such adventures. John has purchased a Practice, which he says will be better than the one in Middlethorpe. He sends his love, but is still very low.

The Leggates' chattels arrived in August. Vans went back and forth. Marian's piano was unpacked. Furniture they had rented was returned. Marian planned a music room on the ground floor. She supervised the instrument's unpacking, urged care as it was moved into position. It was undamaged, out of tune from salt sea air, from hoisting and shoving, from yawing, pitching, and rolling across the Atlantic fastness. She sent out the maid, who brought back the tuner whom she had already found. That evening she spent two hours with her piano.

Two extra women were hired to clean everything, arrange, tidy. Within a few days the house was, for the most part, straight. Dr and Mrs Leggate, of Queen Street, Newborough, were able to dine from their own plates. Marian had felt in almost constant motion since Frederic recovered. Now she could breathe.

"I may give lessons," she said. "Mrs McNair said she would like her daughter to take up the piano."

"You mean for younger children?"

"For girls of perhaps thirteen and above."

Then after a pause she said: "And I think a town like this would do well with a conservatory of music," Marian said.

"For teaching or performances?"

"I shall write to Leipzig, and ask them to send me a history of the founding of their conservatory. It would help to know what's involved."

"Raising money, I expect, perhaps attracting some famous musician," said Leggate.

"There are none here, of course. Perhaps we could get someone in from New York, for a few months."

"You could give performances."

"Should you like me to do that?" she asked. "Mrs Marian Leggate, wife of Dr John Leggate of Queen Street, will give a recital of works for the Pianoforté."

"You think my practice will need the help of extra advertisement on the playbills."

Marian looked at her husband, unable to judge whether his remark were affectionate banter of the kind she liked, or if he were being sarcastic.

"Oh, John . . . I think it will be all right here," she said. "Together."

Chapter Sixty-three

Marian wrote to her mother:

> I miss you very much, Mamma, and wonder how you are, and whether you
> are getting on all right. I try to imagine your days, now everything has
> changed so much from the house in which we all spent so long. You said
> you could never be doing with a Colony, but really this is not bad, and if
> you could just manage the Journey, which as I told you isn't really too
> uncomfortable, you could see us here, and seeing us, be able to think of us,
> and our lives, as I try to think of yours.

Marian paused in her writing. Coming here, she thought, John seemed to
think there was no alternative, but did we think enough? My mother will
never come, she becomes seasick almost if she stands on the pier. But I could
visit her, perhaps next year, or the year after . . .

Leggate would also write to his mother. In January he wrote:

> Snow is a foot deep, and we bought a sledge second-hand, to be pulled
> by the horse. Little Frederic loves to ride in it. On days when the weather
> is about freezing point, which here is thought very tolerable, we wrap him
> up warmly, and nurse comes with us, as I make my rounds. The horse
> has more to pull, but there are no steep hills. When I go in to a patient,
> the two of them may build a snowman, or if the little fellow has be-
> come chilled and I know the people, he and the nurse come inside for a
> hot drink.

The Leggates made the acquaintance of the Ormerods, who in due course came
to dinner and were pleasant. Professor Ormerod had an important position

in the Medical School. John mentioned his researches and also Snow's. The Professor said his experiments were uncommonly ingenious.

"I didn't know of Snow's research," he said to Leggate. "But your discovery is far in advance of it. Of course you must write it up."

Leggate was invited to join the newly formed Newborough Medical Society, to address them, not as any kind of trial by ordeal, but as an honour.

Marian wrote to Caroline of how she persuaded her husband to prepare his address.

> He still thinks I will find fault with it, but I persuaded him to tell me what he would say. He will start with the disease coming from abroad, and use Snow's illustration of how the disease follows the high roads, but not the villages off to the side. Then he will talk of his Alimentary Theory, of the sporules that make the fluid leak from the blood vessels. Then he will describe his own researches, starting with the Five Houses in a Line. I made some tracings of it, nurse helped, so we have twenty-five tracings, which he can pass out when he speaks. Each listener will have a little map to take home, with which to remember the Address. I wrote at the top of each: "First firm Evidence found by Dr John Leggate in the autumn of 1847 that Cholera is spread by Contaminated Drinking Water." After the evidence of the 1832 epidemic he will go on to his experiment of supplying the lodging houses with spring water.

A later letter from Marian told Caroline how the address was enthusiastically received, her husband quite fêted. As a man who had worked with Magendie, he was asked to give a lecture on principles of experimental method, and one on the circulation of the blood, in the new medical school. There were plans that he would also give a course of lectures on prevention of infectious diseases.

Marian wrote of the birth of her second child, Anna, and of her plans for the conservatory of music. She formed a committee with a small group of teachers of the piano and violin. They arranged concerts. She started raising money for the conservatory building. She wrote, too, of recitals in which she played, one attended by fifty-three people, one by eighty.

And there were journeys.

> They are building railways like mad. We went by train to Charville, a bit more than two hours distant on the train. At first we went across arable land with farms well kept, but then one comes to the Primaeval Forest, punctuated by lakes. Seeing it is like being the first person in the world.

Marian received Caroline's letters, with news from Middlethorpe. An incident that agitated the town concerned one Harry Slestor, arrested for so grievously wounding a man in a fight that the man had almost died. In court it was shown that Mr Slestor lived off the earnings of women, to whom in return he offered services of a dubious kind. The most shocking revelation was that when the police had searched his substantial house, they found accounts of monies from women, among whom was Marie Watson—the woman who had been found in front of St Michael's with a dagger in her heart.

Though the case was not as sensational as that in which Edward Warrinder had made his name, it threatened at one point to attain that level of interest when the prosecutor painted a picture of Harry Slestor as a man given to demonstrative actions intended to cause alarm to the women who were under his power. When he came to the subject of Marie Watson, he questioned Slestor closely about her, and asked whether he had not left her with a dagger in her heart as a warning to other women. Slestor was visibly rattled by the insistent questioning. He denied the crime, but there was uproar in the court when, asked finally by the prosecutor whether he would swear to God that he had nothing to do with killing Marie Watson, he shouted: "I did know her, but I swear I didn't have nothing to do with giving her no arsenic, or anything, or stabbing her."

The new newspaper, *The Middlethorpe and District Mail*, which specialized in court reports, pounced on the incident, and asked why Slestor had mentioned arsenic as a poison, when this substance had never been mentioned at the trial of Biggs. Why would Slestor mention arsenic? Could there be any other reason than a guilty conscience? Did not this trial suggest that the murderer of Marie Watson may not have been Biggs but Slestor? So had not the special correspondent of *The Middlethorpe Examiner* overreached himself? Was the conclusion, for which that newspaper had worked so energetically, to have an innocent man hanged?

Caroline wrote:

I saw Warrinder—did you know he is married, to Martha Applegate? Anyway I saw him, and he was terribly exercised about Biggs. He said: "After the hanging I couldn't get him out of my mind for a long time, twitching at the end of a rope, and thinking what if we were wrong?"

I told him that it was the jury who sentenced him. He hates being held up to ridicule by the other newspaper, but does it not speak well of him that he can feel doubt about his actions?

Leggate said to Marian, "I'm invited to write something for the *Upper Canada Medical-Chirurgical and Scientific Journal.*"

"Is that an honour?"

"It's been recently formed. It would be an opportunity to publish my article."

"You should do it then."

"Would you help me?"

"You're sure?"

Marian read the paper he had given to the Newborough Medical Society, rewrote it somewhat, so that Leggate would not tread on the toes of Magendie whose experiments on the injection of mercury he mentioned. But she made it clear that in 1832 Leggate already had the alimentary theory of cholera, and that by the end of 1847 he had shown it was spread by contaminated water. She rearranged some of the matter, incorporated some information from his notes, rewrote paragraphs so that the purpose of each was clear in its first sentence, then reorganized the material some more.

When he was to read her draft, Leggate made sure she was out of the way. He read, and he became pensive.

The findings came through admirably, facts from the epidemics of 1832 and 1849, all put forcefully. There was even a mention of prostitutes and the poorest lodging houses, but not in any way to scandalize, merely to show how much safer had been those who drank uncontaminated spring water than those who drank the town water, wherever and however they lived. He read his idea of the cholera sporules able to live in water supplies, and to multiply by the million in the human intestine. He read of precautions of clean water, of disposal of waste, and of the importance of washing one's hands, and of the hygienic implications of cooking. He read the conclusion that the alimentary theory of cholera was not only better substantiated than other theories, but was corroborated in the work of Dr Snow.

"You have rewritten it a good deal," he said.

"But preserved your sense, I hope." Marian eyed him apprehensively. "Even included qualifications where these were necessary."

"Improved it substantially," he said.

"You don't mind what I've done?"

"I only wish I could write like you."

"You have me, you don't need to be me."

"I write notes every day, but for publication it deserts me. I read it and wince. When I try again, it becomes worse."

"But your thoughts, they're all here."

"I should be able to do it myself."

Marian pressed herself against him. "Can't you see? It gives me some part in it."

When the article was published, Leggate read it with pride.

Not bad, he thought. The journal's circulation isn't large, but it'll have an influence, it takes a while of course, but it will, and it'll be in the libraries, and when the truth is widely known, and when histories are written, my discovery and Snow's will be there.

Leggate said to Marian: "Here's my article . . . our article, that you helped write."

"Let me see." She read the first paragraphs silently.

"In print it looks more official, don't you think?"

"It's real," she said. "We must celebrate."

"You don't mind the mention of prostitutes?"

"What I don't like is being excluded. I was jealous because of something in which you wouldn't include me. When we had those talks when I had my collapse, I thought that would be the end of it."

"Such things are hard to root out."

"It's not that," she said. "It's that you and I, between us, hadn't reached the state that would prevent them."

"And now we have?"

"Now that you're not so abstracted, and can allow me nearer, so I can help, we may have."

"And achieved your aim."

"My aim?"

"A marriage of friendship, in the first article you wrote, that you also showed to Warrinder."

"Not achieved, but approached."

Leggate started to have an influence. It extended some way beyond Newborough, into other towns of Upper Canada. At the university he became respected, and was given an appointment.

When in 1854 a further wave of epidemic cholera reached Newborough, Leggate was not merely established, but listened to.

After the epidemic passed, Marian sent clippings from the town newspapers to her mother, to her aunt, and to Caroline. The clippings were full of praise of Leggate for his public spirit. By a good stroke his recommendations as to the supply of clean water were, in one quarter, accepted. He had persuaded one of the water companies to draw water from a deep well during the epidemic, in preference to the lake into which all the sewage was poured. The virtual absence of deaths in the area supplied by this company was compared with the deaths in an adjacent area, which were many, although that area was more prosperous. The difference was noticed. Leggate's recommendations for the treatment of the sick by hydration were also successful.

By the end of the epidemic his name was on the lips of those who cared about the health of the town. His practice grew, and he had to appoint both an assistant and an apprentice. In his teaching at the medical school, his unconventional and thoughtful analyses made him popular. He had found a way of transmitting his ideas so that they might make a difference.

A further child was born, named Sebastian after Herr Hoffman, and the Leggates felt that at last they had begun to prosper.

Marian gave more recitals. On some occasions she played with a string quartet—their playing of Schubert's Trout Quintet became a favourite in the town—or she would accompany singers. Her teaching of the piano became known; as well as earning money, something of a novelty for women above the class of governess, she could enjoy passing on to others some of what she herself had learned.

Chapter Sixty-four

In 1855 Marian returned to Middlethorpe for her mother's remarriage. The Leggates thought the journey would be too much for small children. They thought that not just a nurse but a parent should stay at home, so Marian went by herself. Travelling alone was becoming more common for women.

"Will being married again make you happy, Mamma?" said Marian at their first intimate talk.

"One doesn't just marry for happiness," her mother said.

"I was meaning in an old-fashioned sense, of living for the best, and with one's heart."

"Then it will," her mother said. "He is a good man, and you are to be generous to him."

"Now you've said this to me, I shall."

"Your recitals in Canada, are they going well?" asked Caroline, eager to catch up with news.

"Not frequent, but they go quite well. I even have a small following." She looked apologetic. "May I come and play your piano? I played a little on the steamer, coming over, to keep my hand in."

"Of course . . . how do you manage, being separated from your children?"

"I shouldn't have left them. But they're well enough with John and the nurse. I'm worried about Sebastian, he is painfully shy. Frederic was different, and so was Anna—she's very confident—but I can't bear to see it when Sebastian clings as if his life depended on it, he's nearly two now and it only seems to get worse."

"Children are shy at first."

"This seems more extreme."

They discussed the problem, discussed Frederic's health for, after the child's illness, Marian and her husband had both worried that he remained delicate. They discussed Anna's ready taking to conversation. And they talked of Caroline's children, and her husband, and their pursuits.

With her aunt, Miss Mitchell, Marian spoke of her accomplishments in Newborough, of her concerts and her work for a conservatory of music. She spoke too of her husband who was respected now.

"I should think he is respected," said the aunt. "And you as well, I've no doubt."

Marian looked at her, noticed her face no longer youthful.

"Respect's important," continued the aunt. "But what about the really important things?"

"Which are?"

"The two of you, you and your husband?"

The question took Marian by surprise, as if confronted by something not admitted to herself.

"I think we are all right," she said. "I think so, although coming away like this, and having all this time to reflect, has made me think . . . perhaps made me think too much."

Mrs Brooks's wedding day arrived. Marian and William duly saw their mother become Mrs Dysart.

At the wedding breakfast at the house of her new stepfather, Marian was surprised to see Warrinder.

"Marian!"

"Whatever are you doing here?"

"I heard you were back."

"How dare you come here? . . . after what you did to my family."

"An hour ago you and I became members of the same family. We've become cousins, or something of the sort."

"I hope you behave better as a cousin than you did as a friend."

"I wanted to see you, to apologize. I was wrong about your husband."

"Is this brought on by a surfeit of family feeling? You should say what you have to say directly to him."

"Please, you don't understand. When I wrote that article, I was wrong. I didn't know at the time . . . I don't know what to say."

"You were obnoxious."

Warrinder could not meet her eyes. "I did your husband and your family an injury."

"I hear you are married now," she said.

"I am," he said. He made a gesture towards the former Miss Applegate, who was standing a little way off. "You know Martha."

"I shall go and congratulate her," she said, unable to keep the note of sarcasm from her voice. "So, has marriage brought on this repentant attitude? Come over here to the writing desk. I'll give you our address in Canada. You can write to my husband."

"You see . . ." he looked sideways at her, "I've received a report just published by a Dr Snow in London. It shows conclusively that water supply is the cause of cholera, just as your husband found six years ago. By rights, your husband should have had priority, someone in this town . . . a discovery . . ." Warrinder trailed off into silence.

"As editor of Middlethorpe's newspaper you lost an opportunity to trumpet for the town. Why not write an article now, explaining how one of Middlethorpe's citizens made a discovery. You could say *The Examiner*'s editor was too benighted to see it, and instead hindered the discoverer."

Warrinder was silent.

"When you've written it, send a clipping to my husband."

More slowly than was necessary, she found paper in the writing desk, looked in the drawers of the desk to find a pen, discovered that the nib was unsatisfactory, searched for another, found one.

"Please, Mr Warrinder, could you change this nib," she said.

He did so. She wrote the address. She handed it to Warrinder.

"I bid you good day," she said.

In one of his letters, Leggate wrote to his wife:

Frederic had a very good thought today. We were at the docks, looking at the steamers. I said the shipbuilders tried to make the ships as fast as possible, to get things to people as quickly as could be. I was explaining that being low in the water made the ship go slower, because it had to push through more water, and that slowed it down. "Why does not someone make it higher then," he said, "to make it go faster." I love it when he takes an interest in how things work.

When little Anna holds my hand as we walk along, I feel it full of trust, as if she can hold on to me for ever. Not just her, the others too, a little hand in mine. They will never remember this, the hand in hand as we walk, but I will, always.

Poor Sebastian continues unbearably shy; I feel for him. I am planning ways to help him overcome it.

Lord Andervilliers heard that Marian was in Middlethorpe, and had his secretary write to her: "Would you consider coming to play for us, at a musical evening."

"You should go," said Caroline. "Why not?"

"Because my playing is rusty."

But she went—there was the same grand house, with its same beautiful piano. Lord Andervilliers and Lady Andervilliers were equally courteous. For the occasion there was a poet down from London, a painter who lived in York, a cellist from Germany, and—most extraordinary—a flautist from Prague, Tereza Czermak, slim, fair-haired—the embodiment, so Marian thought, of feminine perfection.

Towards the end of dinner Lord Andervilliers made a proposal: "Perhaps we should each give a little speech on some subject, as they once did at the *Symposium*, in Athens."

"And not hear the flute girl? I should oppose any such scheme," said the cellist. "I've been looking forward to hearing her." He bowed slightly towards Miss Czermak.

"Not instead of music, but in anticipation of it," replied Andervilliers, and then explained the meaning of what the cellist had said. "You see, each person at Plato's party had to give a speech in praise of love, and because they decided on the speeches they sent away the flute girl who was to have played."

"So you will speak first," said the poet.

"Why not one of the ladies?" Andervilliers turned to his wife. "My wife always has something novel to say. What about the subject of love—to shake our prejudices?"

"I have nothing to say. I am the most conventional of creatures," said Lady Andervilliers. "Love seems to me too often an excuse for impropriety." There was a ripple of laughter.

"I told you she'd say something striking."

"In the *Symposium*, I like the speech by Aristophanes," said the poet, "in which he says that each human being, as we are now, is only half a person— we used to have four arms, four legs, two heads and so forth, but were cut in half by the gods for our arrogance. Love is the residue, yearning for our lost half."

"That would make love the cause of everything," said Lady Andervilliers.

"We could begin where Socrates left off," said the cellist. "He thought that love between people was a wretched thing. We have to start with it, but then must transform it to something higher, like the love of beauty or knowledge. Of knowledge, I know nothing." He paused to see if anyone would appreciate his joke. "But beauty, perhaps, such as music."

"To me it is quite the other way round," said Marian. "I would start with something like the love of music, which is perfect. The trouble comes when one tries to apply love to earthly concerns."

People smiled appreciatively at her joke.

"The actual falls so far short of the ideal," said Miss Czermak, clearly and suddenly. She had not spoken before, but now she spoke in such a meaningful way that the silence that came afterwards was palpable.

To Marian the utterance of this woman seemed abruptly to reveal such a passionate devotion infused with such an intense disappointment that her soul seemed about to crack.

Marian rescued her from the silence. "Ideals do have a way of torturing us," she said.

She saw that no one else was about to speak, and continued. "I know it's not the thing to say, and that Plato makes him ridiculous, but I like the doctor in the *Symposium*, or his idea at least, that harmony is an expression of a certain kind of love."

"You're saying so because you think the medical profession should be commended," said Lady Andervilliers, joining Marian in starting up the conversation again.

"Her Ladyship says that because my husband is a physician," Marian explained. "But I am afraid I can't praise them in any general way. Doctors include as many of the self-important and ignorant as other professions."

People round the table smiled again.

"I meant something else," Marian continued. "The doctor in the *Symposium* said that something like love animates the whole body. But he says there are two kinds. When the elements—hot and cold and dry and wet—are attracted to each other and combine from a good kind of love, the body thrives. He likens this to the harmonies of music. But if the elements love each other in a bad coercive kind of way, that's when the body suffers disease."

"And does feeling the right kind of love for one's fellows promote the good kind of love within?" asked Lord Andervilliers.

"It's possible," said Marian. "I like the idea of a harmony between ourselves and our world, as well as within us—it's not so much that it's an ideal that is beyond us, as Plato always makes me feel, more that it's a state that sometimes one discovers in ordinary relations."

"A most fitting conclusion," said Andervilliers. "Time to approach perfection. Let's hear the musicians."

A gentleman sang Schubert *lieder* to piano accompaniment by one of the ladies. Then it was Marian's turn. She took her Chopin from the music case that she still used, the present from her courtship days that she had brought with her. She scarcely needed the music, but felt reassured by its presence. She played the Chopin Berceuse, then the Nocturne Number 3.

When she finished, she rose, curtseyed to warm applause, was about to gather her music when Lady Andervilliers said: "Leave it, it's the trio next, they won't need the piano." They went to sit.

The cellist and flautist, with the lady accompanist now playing harpsichord continuo, performed a movement from Bach's G-minor Sonata. With the relief of herself having played well despite the disruptions of travel, and with what she had said about harmony still in her mind, Marian found herself swept up into the sonata.

Next, as the flautist returned to her seat in the audience, the cellist stood up to announce two movements from a Bach cello suite.

"These were written when the composer was in Leipzig," he said, "but really for practice, like études, rather than for performance. They're not much performed, but I thought these might be to your liking: the Prélude, then the Fifth Movement with its two Bourrées from Bach's Third Cello Suite."

From Leipzig, she thought—the cello, the cellist, Andrei! What if he had asked me to marry him? I would have accepted, I would, of course I would, but his disappointment in me would only have been postponed. She sighed inwardly. No, she thought, it wouldn't have done...

The cellist had seated himself again, and began to play.

This music, from Leipzig? she thought. Why haven't I heard it? Exquisite, like breathing, drawing energy, then releasing it. Chopin uses this. He studied Bach, must have studied this. All texture, waves...

When the Prélude ended, the Bourrée began, a joyful dancing melody.

Then the flautist, Tereza Czermak, came forward again, and played unaccompanied: Bach's Sarabande from the Partita in A Minor. Marian found herself transported back to those ancient times of the Greeks, whom they had been discussing.

In an orange grove outside Ephesus, she thought. At the Temple of Artemis. The flute sounding through the trees. It's the goddess herself who is playing, wearing a garment of the lightest silk, which wafts gently in a cooling breeze. We mortals approach the temple, mount its marble steps, up under the great parapet, within the columned hall, to the sacred area, and there she is...

Whether it was the sound of the flute, or the sight of Miss Czermak's feminine perfection, or the combination of the two, Marian felt that something extraordinary had happened, as it had sometimes when she played Chopin, or here as if a goddess really had come down to earth for a brief moment.

The piece ended, people applauded, and the mood was broken. The musical part of the evening ended.

Mechanically, Marian went to the piano, opened the stool in which she had placed her music case. She gathered her music books, placed them in the case. When she looked up, she realized she had excluded herself from the knots of conversation that had formed, the largest of which was round Miss Czermak.

Marian felt disinclined to join the others... she put her music case beside the piano, and walked to the library, the door of which was open.

So many books, she thought. Is this what I should have aimed for, tried to marry someone with all this?

She gazed along the shelves. Look here—Lucretius, she thought. What a lovely binding.

She took the book out, and started to leaf through it.

"Can't we interest you in conversation." Lord Andervilliers startled her.

"I'm sorry, I hope you don't mind me coming into your library. I love books."

"Not at all, please, just that I didn't want you to feel left out ... found something good?"

"*De Rerum Natura—On the nature of things.*"

"You read Latin, you know Plato, you play Chopin—you're a remarkable woman."

"Not I, the remarkable woman's in there—this is the piece I was looking for."

"You'll need to translate, my Latin's too rusty for words."

"Something like this," she said:

> "... he thinks himself a fool
> That even for a moment he'd imagined she was blessed
> With charms we mortals cannot ever have."

"So that's what you women think of each other."

"Not generally—but doesn't she seem like a goddess? Wouldn't any man find inspiration with such a muse? Isn't it hard for us all to avoid going to join her circle of worshippers?"

A few days after the musical evening, Marian went to stay with her brother at Hadby Grange.

"I hear you were with my neighbour Andervilliers," said William.

"Your neighbour? He must be at least ten miles away."

"Neighbour in a manner of speaking," said William. "With an estate, I'm in the same line of trade."

He smiled, and his sister thought how much he had changed from when they had both lived in their father's house.

"You're well set up," she said.

"It suits me."

"So it seems. And what do you think of Warrinder becoming a member of our family?"

"Not much we can do about it."

"He told me at our mother's wedding that he was repentant about that scurrilous article he printed about John. Did you say anything to him about it?"

"Nothing whatever."

"You're sure?" Marian regarded her brother, looking him in the eye, until he turned his eyes away.

"You did say something," she said.

"Why would you think so?"

"I know you."

"I've not said anything," he said. And then after a pause, unable to resist his sister, he smirked and said, "That's not how these things work."

Marian remembered herself uttering just such a phrase to her husband when he'd questioned her about seeing Gadd for his application to the Medical Society. She looked again at her brother, pursed her lips, and scowled at him.

"So now you're a man of property you can go about giving a nod here, and a wink there."

"Only in a good cause."

"Did you find marriage what you expected?" said Marian to Caroline, a few days before she was due to return to Canada.

"I didn't have such strong expectations as you."

"You know what I mean."

"There are some disappointments—Peter's wrapped up in himself too often, and doesn't always take much notice of the family," she said. "I've become more of a mother than a wife."

"And do you mind that?"

"It's been worse without you here. I miss you, and don't any longer have anything like our friendship, though my Emily's a companion now, she's so nearly grown up, and lovely to be with and talk to."

"So good things that weren't exactly expected."

"You seem preoccupied," said Caroline. "Something's on your mind."

"It's so strange coming back here, I find it hard to think of my mother marrying again, and then there was Warrinder wanting to make an apology to John, and my brother being the lord of the manor, and everything else. So much has changed. And I miss you more than I can say . . . and in a few days I'll be on my way back again."

Marian paused, not knowing how to say what was in her mind. "And . . . at Andervilliers's the other day," she said at last, "it stirred me up. Life here, among those kinds of people at any rate, is so settled, so secure. Quite opposite to Canada, where everything seems more fluid . . ."

"Something you're apprehensive about going home to?" asked Caroline. "To John."

"No, no, not that. Well, not really . . . It's just that at Andervilliers's there was talk after dinner about the nature of love, someone talked of Aristophanes's idea of people having been cut in half by the gods, long ago, and being driven to find their other half. I don't know why it affected me so. Perhaps because it's so utterly different from the ideas I used to have of friendship."

"You were always very theoretical."

"And the idea of a theory is to have it tested against the world," said Marian. "For the theory to get better—perhaps that's happened. John and I didn't really have the right theory to start with, about what kinds of parts of each other we should be."

Marian saw Caroline looking at her. Perplexed, waiting to hear what she would say next.

"There was a flautist there," Marian continued, "who was quite extraordinarily beautiful, like a Greek goddess, feminine, the way I have always thought a woman should be, somewhat like you in fact."

"You've not always thought of me as the way a woman ought to be."

Marian looked at her friend, surprised. "Indeed I have."

"Then you're very silly."

"Because you're right, I am too theoretical. Perhaps what John needed was someone more like that, a muse, not someone to help him in the construction of sentences. I worry a good deal about whether I'm what a man needs—what John needs."

On the voyage home, Marian would come from her cabin onto the deck to feel the Atlantic wind on her face. In spells of calm weather she sat near the leeward paddle-wheel, hearing its thudding, gazing across the infinite sea.

When not gazing at the horizon, or watching the paddle-wheel thrust into the hard green water, she read Madame Sand's novel *Les Maîtres Sonneurs*, about music and love, written three years previously, in the wake of the author's grief for Chopin and disappointment in the 1848 Revolution.

The book's about love as a kind of religion, she thought. To achieve that passionate ecstasy is to be pure. But it's not enough. No better than worshipping a goddess. The same impulse in fact.

Chapter Sixty-five

In his study in Queen Street, Leggate jotted down preliminary ideas for the new supplies of water in Newborough—the question was whether to adopt the obvious solution and take water from the lake, but with the intake far out in deep water and some miles to the east of the town, or whether to take water from one of the rivers, or from some other source. He and Professor Ormerod were to discuss the matter with the mayor.

The doorbell rang, and Leggate, going ahead of the maid, went to answer it.

As he opened the door, he blinked, for the outside light was bright—it was Marian, back earlier than expected.

Both shy, after so long an absence, they embraced.

She had taken the new steamship route to Montreal, then changed steamers and coming on directly, she had overtaken her letter. From the steamer quay, she had come by cab straight to the house. Relieved to see her children and her husband safe, she distributed presents, told them of her adventures.

"You should have sent a telegram, and we would all have come to meet you."

"I thought I'd surprise you."

When the children were in bed, Marian and Leggate sat together.

She said, "I'm glad we've moved here."

"I've come to like it too," he said.

"You don't miss England?"

"I feel more in place here, feel I can do more, or in a way that suits my talents. But you? Don't you miss England?"

"I've thought about it a great deal . . . before I came home, I went to Lord Andervilliers's house again—I wrote to you about it but probably the letter's not arrived yet. That would have been my life there, soirées, the occasional concert party, playing at rich people's houses—that's what musicians do in England. I can't imagine a conservatory in Middlethorpe, but here . . ."

The two sat in silence for a little while.

"Warrinder wrote to me," Leggate said. "He sent a copy of his article, and made an apology."

"So he should."

"Did you say anything to him about your brother owning his offices?"

"Certainly not."

"Then your brother said something?"

"I asked him directly, and he said not."

"Hmmf," said Leggate.

"The woman whom Warrinder married is related to my mother's new husband," said Marian. "Perhaps becoming a member of the family has given him a qualm of conscience."

"Anyway, they've pulled down those houses in Friars Walk, did you know they were doing that? and they're building a Public Library there. Warrinder says that, in the vestibule, on the wall opposite the plaque on which will be written the names of those who have subscribed for the building, will be one with my name on it."

"You subscribed?"

"No, on the opposite wall. Warrinder says he's proposed that the plaque should have an inscription. I'll read you what he proposes."

He went to his study to retrieve the letter: "On this site stood a lodging house, the inhabitants of which were protected from the cholera epidemic that ravaged the town in 1849, by drinking only pure water drawn from Tafts Spring, as recommended by Dr John Leggate, who lived and worked in Middlethorpe from 1845 to 1850. He was the first to show that cholera is caused by drinking contaminated water, a discovery that provides the basis for the most important advance in public health yet made."

"He has the grace to admit he was wrong," said Marian.

"I should have liked to give you something more."

"It's enough for me. Is it enough for you?"

"It's not what I hoped for, but the Middlethorpe epidemic seems so far in the past now. Things are changing. People are thinking more clearly about public health—while you were away I was asked to consult about the town's new waterworks."

"So, more than most people, you've been able to help make the world a better place—it's a privilege."

"It is, and it is enough."

Though married now for seven years, Leggate and Marian felt strange to each other, after the long separation.

"I've not slept well, while I've been away," she said. "On the way home, I found it difficult to sleep at all."

As on their first night together, she wanted to let him feel her presence, until the strangeness, like dew on a warm day, evaporated . . . he felt a slight twitch of her body as she fell asleep. Leggate put aside a certain disappointment. Resting his hand on her head, he thought about the house in Friars Walk, his article, his plaque . . .

At breakfast next day, Leggate said, "Since you've been away I've been taking Sebastian out with me on fine afternoons, thinking how to help him in his shyness."

"You agree there is something."

"I feel for the little fellow. Frederic and Anna and Sebastian and I go down to the docks. Frederic likes to watch the steamers, and I have given him a fishing rod, and Anna brings a doll. With Sebastian I have an agreement. If he will say 'hello' when we come up to another child, then he may ride on my shoulders."

"And this is having an effect?"

"With children, we do quite well. With adults the idea is that if I say 'how do you do' to someone, he must say 'hello,' not cling to my leg and hide behind it."

"But you have not progressed so far?"

"We shall persevere."

That evening, when the children were in bed, she said, "We could choose a novel, and read aloud to each other."

"We could."

"You choose then."

"Something ironical, Jane Austen?"

Marian rose, a minute or two later returned with a book. She positioned herself close to a lamp. "It is a truth universally acknowledged, that a single man in possession of a good fortune must be in want of a wife..." she began. "Do you know I was reading Jane Austen when you attended my aunt, when we met after the piano competition?" she said.

"I'd forgotten."

"You're hopeless."

"Now I'm the invalid and you'll read to me?" he said.

"We'll take turns to read the chapters, and we'll discuss each one. I shall tell you what they are feeling, and you ... you can tell me what they should do next."

"Read the first chapter then, and afterwards why don't we go early to bed?"

"Too ironic for now, we'll start the book tomorrow."

In bed: "I'd wanted to give you something, a sort of offering ... to deserve being looked after by you," he said, as he cradled her head on his breast. "More than a plaque on a wall."

"Like a boy giving something to his mummy, so that she'll love him."

Leggate laughed. "Why put it like that?"

"I don't have a special feminine stuff to give you," she said. "You should know that's just a kind of illusion that women give off, some women anyway, a kind of magic."

"It seems to me you have that stuff."

"But you speak with anguish about how patients want you to be a kind of parent who can make it all better in some magical way."

"That's not the same kind of thing at all."

"Is it not?" she raised an eyebrow, quizzically, smilingly. "Well, I like you looking after me. I've missed you."

"So equality is harder to achieve?"

"I didn't know that when I wrote my article on friendship, I had no idea of the real nature of things. I thought friendship should come first, and be the only thing."

"But now?"

"There are the longings too, less easy to understand, that work within and between people. It's not so much a matter of starting with friendship, or imagining ourselves complete in ourselves, as coming through the longings to friendship."

"And we have?"

"We can be closer . . . now. I don't have the yearning I used to feel so strongly. I did find what I wanted, found it in you, I mean your kindness, your moral passion for science, to make the world a better place."

He looked at her appreciatively.

"And me?" he said. "Have you discovered what I've found in you?"

"I used to think it was respect you wanted," she said. "But now I think that was only a substitute. You wanted to regain a love you'd lost."

He looked at her again. "Yearning and actuality do match in you," he said. She drew him closer.

"But must we achieve equality here in bed?" he asked. "I'm not sure I can do without that sense of receiving from you something I don't have, a tenderness . . . your love."

"I like your tenderness too," she said. "Perhaps, so long as it's a secret, we could keep it, for here sometimes, a vestige of something past.

"It was just after I first met you, when you attended my father, that I first started to feel, sometimes when I played Chopin, that I could achieve a kind of state, of being outside myself. And then, when we were engaged, I felt it sometimes with you. I suppose that's a kind of magic as well. And sometimes it has happened here in bed, but I don't so much want to be outside myself any more, just here within myself and with you."

They kissed, feeling each other's presence more fully now, physically, mentally.

"I'm glad you're home," he said. "After the long journey."

"Are you home too?"

"I am."

Chapter Sixty-six

It was five years after his wife's return from England, and Leggate was bending over his map of the district of Newborough: the harbour, the hospital, the theatre, the principal church, the proposed site for the new waterworks, agreed now after long delays, with the work due to start soon.

Though where best to put the waste? he thought. That's still not solved. Or perhaps we could rid the world of disease sporules entirely. Not just the little plants and sporules of cholera, but the others, typhoid, diphtheria, the other fevers, all of them.

He had an impulse to talk to his wife, to have a cup of tea with her, perhaps. He went to look for her.

"If we could understand their natural history," he said, "understand how sporules of all the different diseases propagate, then we'd know better how to plan everything."

A few days previously, Marian had received a parcel of new books from London. She had been reading one of them. Hearing him enter the room, she looked towards him, to hear his pronouncement.

"Some things are more difficult to plan," she said.

"But it's necessary to plan waterworks, what matters is whether we do it right or not."

"I've finished the book that everyone in England's so excited about," she said, *"On the Origin of Species."*

"Species of what?"

"All plants and animals. You should read it, it's very significant for your theory," she said.

She held the book up.

"We're all competing," she said, "every species, with each other. You'll need to plan for species of planticle that are not purely passive."

"You mean they'll fight back?"

"Darwin says we compete with the planticles, and they with us. You know

I used to be interested in Malthus? He's the key—he's shown that people multiply in proportion to the number already present, the more there are, the larger the product with each new generation."

Marian looked at her husband to see if he were following.

I wish he'd read more, she thought. Different things, open his mind up, instead of just concentrating on disease all the time.

"Malthus?" said her husband.

"If we could avoid swallowing or breathing all the sporules of disease, we would win the competition with them, and people wouldn't die so much, but then a man and wife who had six children would multiply themselves by three; then each of their children would multiply by three, and so on. Without disease, in a hundred years a hundred married people become eight thousand, in a thousand years not enough room on earth for us all to stand up. I've been making notes and calculations."

"So you forbid us to work to stop disease?"

"I'm saying that now we know how to avoid sporules, we must avoid multiplying as well, or everything will go out of balance."

"I say we will have no more marriage! Is that it?"

"We can marry, but not multiply—we need to understand the natural history of marriage, not just of disease. Food would run out, and human misery would multiply, far more than at present because there would be more people, and that's when the planticles would reassert themselves because they need less, and can just lie dormant."

Leggate raised an eyebrow.

"Your work," she said, "the measures to stop transmission of disease, they'll be salvation for the individual, but if we don't stop multiplying, they'll be disaster for the human species."

Next day, Marian was still reflecting on Darwin, and on the inexorable law of Malthus.

"Darwin's book was published at the end of last year," she said to her husband. "So, in or about December 1859, human character changed."

"With the publication of a book?"

"You've not read it. It starts a new era. We've all assumed—more or less—

that we were made in the image of God, and are fallen creatures. Most people do fall—into wretchedness, and never rise again. Only a tiny few make their way back to become reconciled with God. Then what have they achieved? They've arrived back to where they started."

"Reunited with God is not bad."

"That's the way Saint Augustine said we should each see our own life—fallen so the only question for each human character, each human biography, is whether we can repent and become one with God again. But it's changed."

"So what now?"

"We've evolved from animals, each species changing by infinitesimal gradation into another, in the first instance from the meanest animalcules. That is our new story. We are not fallen, we're making our way up from something worse."

Leggate remembered Amy saying something similar when they'd been in Middlethorpe, but thought he would not mention it.

"Did the prostitutes you used to know stop themselves conceiving?"

He looked alarmed. Must be chance, he thought, she couldn't have read my mind.

"I think they did," he said.

"I've been thinking," she said. "I went to the library this afternoon, while you were on your calls. Malthus was wrong about one thing; there are other means for keeping down the population than abstention, or death by famine and disease. We could prevent ourselves multiplying by using these means. I talked to the librarian, he recommended a book—*Moral Physiology*, by the son of Robert Owen, you remember I wanted to visit the place of his experiment, in Scotland. The son writes about preventing conception."

More difficult than discovering microbes, she thought. Discovering how to persuade people against their instincts.

"Preventing conception's been written about for centuries," said Leggate. "It's not popular. People wouldn't do it, they would consider it coarse."

"Not worse than abstinence."

"You think we doctors should go round and prescribe such things, as soon as a couple has two children?"

"We should tread more lightly on the face of the earth. If humankind has evolved so far, so that we can think, then we must think about our actions."

"People scarcely think about reproducing, we just do it."

"Not think then—cooperate, that's what we humans can do. Though the different species compete, we human beings are the creatures who succeed because we do together things that can't be accomplished alone."

She paused as she heard her own words. "Like you and me," she said, "love in a good way, as friends, help complement the other person, not in a coercive way that leads to strife."

"The task for the next era?"

"No need to be sarcastic," said Marian. "You like large problems. Or why not a problem for us both? How the human race can stop multiplying, before we all reproduce ourselves out of existence, so the only creatures to survive will be ones that are too small to be seen."

A Note on the Discovery of the Causes of Infectious Disease

This novel is based on human beings' first serious recognition that we share this planet with micro-organisms—germs—that in us can cause deadly disease.

Cholera was the disease that first made clear that human infections were caused by unseen bacteria and viruses. Although it once seemed that such microbes might be exterminated, we now know this is impossible. Their number, multiple habitats, resilience, and mutability indicate that it is likely that we will die out before they do.

The historical discoverer, whose article in *The London Medical Gazette* of 1849 enters the novel, was John Snow. He was the first to publish the correct (and testable) hypothesis about the alimentary mechanism of cholera and about how the disease spread. He realized that the cholera poison was biological, and that it "multiplied itself by a kind of growth."

Snow's discoveries came in several steps. The final steps came during the third English cholera epidemic of the nineteenth century, which erupted in 1853 and 1854. Snow traced a large number of deaths to a pump in Broad Street, in Soho, London. He stopped further spread of the disease in the area by applying to the guardians of the parish to have the pump handle removed. In *On the Mode of Communication of Cholera* (1855), he gave a map of Soho, indicating the deaths in each house. Its publication cost him £200; only fifty-six copies were sold, earning him £3 12s.

Epidemiology, the name of the method used by Leggate and by Snow, is medical detective work. The Broad Street discovery was one example. But more impressive still, and more conclusive, was Snow's investigation in South London, which he also described in his 1855 publication.

In 1854 the Southwark and Vauxhall Water Company took water from the tidal basin of the Thames at Battersea. London's sewage was poured into this same tidal basin. So twice a day the flood tide brought mixtures of brine and sewage upstream to the Southwark and Vauxhall Company's water intake.

In 1852, the Lambeth Water Company moved its works from Hungerford (in the tidal region) to Thames Ditton upstream of Teddington Lock, above which the Thames is not tidal. Although before this there was nothing to choose between the two companies, from 1852 the Lambeth Company started supplying water that had not been mixed with London's sewage.

From the Registrar General's returns of deaths from cholera in London in 1853, Snow noticed differences in mortality in the areas covered by these two water suppliers. He was alert to the implications. He also had an important piece of luck: he found that during installation of piped water to houses in South London there was an area of Kennington, Walworth, Brixton, and Stockwell in which both suppliers had been in competition, and had piped supplies down almost every street. Particular households took supply from either company according to the whim of the householder. Some 300,000 people, with all the influences of locality, social class, and so forth equal, were therefore distinguished solely by their water supplier.

Snow describes the difficulties he had in obtaining accurate evidence from householders about the sources of their water, since many had no idea which company their water came from. He solved the problem by taking home little vials of water from houses he visited, to analyse them chemically. Water from the Southwark and Vauxhall Company had a higher concentration of sodium chloride than that from the Lambeth Company. In other words the Southwark and Vauxhall water was slightly salty because it was from the tidal region of the Thames with which sea water is mixed, whereas the Lambeth Company supplied fresh water from above the tidal region.

Within the area in which the two water companies' pipes were intermingled, Snow found that between July 8 and August 4, 1854, there were 334 deaths. For 286 of those who died (86 per cent of the total), the water came from the Southwark and Vauxhall Company. For only 14 cases (4 per cent) did the victims' water come from the Lambeth Company—a ratio of 21.5 to 1. Of the remaining cases, 30 (9 per cent) were of people who got their water by dipping a pail into the Thames or into a ditch, or who got water from a pump well. A further 4 (1 per cent) died when travelling, so the source of their water was not known.

These figures do not quite clinch the argument, because the Lambeth

Company supplied fewer houses than the Southwark and Vauxhall Company. By 1856 Snow obtained the figures, and published them in *The Journal of Public Health*, expressing the risk in terms of proportion of deaths to the number of houses supplied. Within the region of overlapping supplies, people taking the Southwark and Vauxhall Company's water during the four-week period at the height of the 1854 epidemic were 6.94 times more likely to have died than those taking the Lambeth Company's water. With modern statistical methods it can be shown that this ratio (6.94: 1) is exceedingly unlikely to have occurred by chance, so something in the Southwark and Vauxhall Company's water must have caused the high rate of deaths.

HULL, LONDON, AND MIDDLETHORPE

Though London could have been the setting for this novel, I decided instead on a town, which I have called Middlethorpe, based on Hull. In the mid-1840s Hull's new waterworks were built, and took water from the tidal River Hull two and three-quarter miles upstream from the estuary. In his account, Snow said that whereas in 1832 there were some 300 deaths from cholera in Hull, in 1849 the number was six times larger (1,806 people), despite improved drainage, and despite 8,000 to 10,000 people leaving the town during the epidemic. We can infer that the increased mortality was largely due to the river having become almost the town's only source of water, so that when the river became contaminated almost everyone was at risk.

In London also, the epidemic was worse than necessary, because neither the nature of cholera nor its mode of transmission was understood. Edwin Chadwick, who founded the General Board of Health, succeeded in 1849 in having London's sewers cleansed by flushing millions of gallons of water through them into the Thames. This measure is discussed by Snow in his 1855 publication. It is likely that the practice contributed to the 1849 death rate from cholera in London being twice that of 1832. Snow suggests two reasons. First, sewer flushing drove cholera germs into the river before they had time to decompose. The sewage entered the river at low tide and was carried for five hours in every twelve upstream towards the water companies' intakes. Second, the practice depleted reservoirs, so that water had less time to settle before being sent to customers.

SNOW AND PASTEUR

In 1858 Snow died of a stroke, aged forty-four. He was not elected to the Royal Society, and even now is not much remembered. Instead, it is Pasteur who is thought of as the great pioneer in this field. He was a pioneer. He showed that putrefaction and infection were both caused by germs. He was a pioneer, too, in that as well as being a great researcher, he was perhaps the first scientist to be successful in convincing a large number of non-scientists of the importance of his findings. In the 1860s he held Paris audiences spellbound with lecture-demonstrations showing that there was no such thing as spontaneous generation, but that putrefaction occurs only in substances that are left open to the air in which, as in water, billions of tiny germs of many species are continuously present.

Pasteur's research on infection was at first on diseases of wine and beer, then on diseases of silkworms. Not until 1876 did he turn to human infections. Nonetheless, it is Pasteur's name that has passed into ordinary language, with "pasteurization," treatment by heating to kill germs in milk and other foods, despite the fact that Snow recommended boiling water for "drinking and culinary purposes" in 1849 and 1855.

So although Snow's evidence was published, it was really not until the 1870s that it became widely accepted that infectious diseases were spread by agents that reproduced themselves—germs. Robert Koch was the first to publish (in 1876) on the isolation of the germs that caused a human infectious disease, anthrax. In 1883 he journeyed with microscope to Alexandria and to Calcutta. From the intestines of those who had died of cholera, and from drinking water, he recovered the comma-shaped bacteria of the species, *vibrio cholerae*. He found how to cultivate them outside the body on beef-broth jelly. He found they were never present in the healthy. In the spring of 1884 he returned home to Berlin, to a hero's welcome: a medal, a vote from the Reichstag of a hundred thousand marks, an official reception at court.

THE NATURE OF CHOLERA

Normally cholera vibrios live harmlessly in river estuaries in many parts of the world, not just in India where cholera epidemics were previously thought to originate. Periodically, conditions of climate and human practices allow the vibrios to enter drinking water, where they can survive for several weeks, especially in warm weather. But conditions for their spread to human intestines are rather specific, and without them epidemics die out.

The main methods of transmission of cholera are by contaminated water, and by people touching clothes soiled with evacuations of victims, and then eating food without washing their hands. Once swallowed, the microbes can multiply rapidly in the human intestine and cause the disease in about forty-eight hours, though normally they are destroyed by stomach acid, so only a proportion of people who swallow the vibrios succumb.

In the seventeenth century, the most famous English physician of that time, Thomas Sydenham, said that "fevers [infectious diseases] with their attendants constitute two-thirds of the diseases to which mankind is liable." On this basis, the discovery of microscopic plantlike germs, each with its own mode of propagation, has been the most important discovery yet made in medicine. Death rates from infectious diseases fell steadily during the last quarter of the nineteenth century and the first half of the twentieth because of improved water supplies and sewage removal, and all this happened before the introduction of antibiotics around the time of World War II. In numerical terms, arrangements in public health have been far more important in the conquest of disease than anything done at the bedside.

Acknowledgements

To be means to be for another, and through the other for oneself. A person has no sovereign internal territory, he is wholly and always on the boundary; looking inside himself, he looks into the eyes of another or with the eyes of another.

Mikhail Bakhtin, *Problems of Dostoyevsky's Poetics*

I would like to express my great gratitude to Jenny Jenkins who, by offering suggestions of plot and character, and by willingly reading six or seven drafts, made this book much better than I could ever have made it on my own.

I warmly thank other friends who read the book and made suggestions that much improved it: Pat Baranek, Ann Craik, Alison Fleming, Sholom Glouberman, Susan Glouberman, Kathleen Jenkins, John Hutcheson, Linda Hutcheson, Debbie Kirshner, Berel Schiff. Thanks to Debbie Kirshner and Susan Hammond for advice on musical matters. My warmest thanks also, for her encouragement and suggestions, to Cynthia Good, my editor and publisher at Penguin Canada as well as to Mary Adachi the copy editor, Jem Bates the production editor, and Martin Gould the cover designer. My agent, Sara Menguc, too, has been helpful throughout.

I wish also to acknowledge some of the sources for my material. First of all I am indebted to George Eliot, whose novel *Middlemarch* was my original inspiration. The events of *Middlemarch* mapped just one path through a world of possibilities. Tertius Lydgate and Dorothea Brooke of Middlemarch, of course, never married, and in other respects are not like John Leggate and Marian Brooks of Middlethorpe, but they were starting points for projecting onto the page the town of Middlethorpe and the lives of some who lived there.

Exceptions to the publisher's note, on p. iv, are that the characters of François Magendie, Clara Schumann, and John Sutherland, who appear in this book in fictionalized form, are based on historical people of the same names. Moreover, Middlethorpe is based on the Yorkshire town of Hull.

My principal information on Magendie has come from the biography written by J.M.D. Olmsted, *François Magendie: Pioneer in Experimental Physiology and Scientific Medicine in XIX Century France* (New York: Schuman, 1944).

Information on Chopin was from standard sources, especially J. Huneker, *Chopin: The Man and His Music* (New York: Dover, 1966). Other background material of piano playing included N.B. Reich, *Clara Schumann: The Artist and the Woman* (Ithaca, NY: Cornell University Press, 1985). Also I benefited from reading A. Fay, *Music Study in Germany* (reprinted, New York: Dover, 1965). Marian's ideas about meaning and music derive, rather loosely, from L.B. Meyer, *Emotion and Meaning in Music* (Chicago: University of Chicago Press, 1956).

Some of the phrases in the court trial of the fictional Augustus Biggs are taken from newspaper accounts of trials of the period or slightly later, by T. Boyle, *Black Swine in the Sewers of Hampstead: Beneath the Surface of Victorian Sensationalism* (New York: Viking Penguin, 1989). The idea of a prostitute writing to a newspaper is taken from E. Royston Pike, *Human Documents of the Victorian Golden Age (1850–1875)* (London: Allen & Unwin, 1967).

For the investigative work on cholera, I have drawn on John Snow's 1849 article in *The London Medical Gazette* of November 1849, pp. 745–55 (mentioned in the novel) and his monograph of 1855, *On the Mode of Communication of Cholera* (London: Churchill), which is most easily available in *Snow on Cholera, being a Reprint of two Papers by John Snow, M.D. together with a Biographical Memoir and an Introduction by W.H. Frost* (New York: The Commonwealth Fund, 1936). I have benefited from the biographical article in that volume, as well as the longer biography by D.A.E. Shephard, *John Snow: Anaesthetist to a Queen and Epidemiologist to a Nation: A Biography* (Cornwall, Prince Edward Island: York Point Publishing, 1995). After writing this book I also discovered a novel had been written with John Snow as its central character, based on his work on cholera during the 1854 epidemic in London, by Liza Pennywitt Taylor, *The Drummer Was the First to Die* (New York: St. Martin's Press, 1992).

Eruptions of the 1848–49 cholera epidemic in Middlethorpe follow those of Hull, as reported by John Sutherland, an Inspector of the General Board of Health (who appears in the novel), in *Report on the Epidemic Cholera of 1848 and 1849 to the General Board of Health, Appendix A* (1850) (London: W. Clowes for Her Majesty's Stationery Office).

For general histories of cholera I have benefited from the standard source, R. Pollitzer, *Cholera* (Geneva: World Health Organization, 1959), as well as from other histories including: N. Longmate, *King Cholera: The Biography of a Disease* (London: Hamish Hamilton, 1966); R.J. Morris, *Cholera 1832* (London: Croom Helm, 1976); M. Durey, *The Return of the Plague: British Society and the Cholera 1831–2* (Dublin: Gill & Macmillan, 1979); W.E. Bynum, *Science and the Practice of Medicine in the Nineteenth Century* (Cambridge: Cambridge University Press, 1994). I have also benefited from consulting general biographies and histories of the period, including H. Mayhew's *London Labour and the London Poor* (1851); J.W. Dodds, *The Age of Paradox: A Biography of England 1841–1851* (London: Gollancz, 1953); and R. Postgate, *The Story of a Year: 1848* (London: Cassell, 1955).

Readers may have noticed that in the last chapter Marian anticipated by some sixty-four years a form of words uttered by Virginia Woolf who said in a paper she gave in 1924 that: "In or about December 1910 human character changed," in "Mr Bennett and Mrs Brown," from V. Woolf, *Collected Essays, Vol. 1* (London: Hogarth Press, 1966). (I have slightly altered Woolf's sentence.)

Epigraphs, and the line drawings that open each part, are in the public domain, or permission has been sought to reproduce them. The author and publisher are grateful to those who gave permission. Despite our best efforts we have not been able to reach some copyright owners, but to these people we are also grateful. Sources are as follows.

EPIGRAPHS

A. Dru. *The Journals of Søren Kierkegaard.* Oxford: Geoffrey Cumberlege, Oxford University Press, 1938, p. 127.

Ecclesiastes. The Bible. King James's Version.

J.M.D. Olmsted. *François Magendie: Pioneer in Experimental Physiology and Scientific Medicine in XIX Century France.* New York: Schuman's, 1944, pp. 183–84.

N.B. Reich. *Clara Schumann: The Artist and the Woman*. Ithaca, NY: Cornell University Press, 1985, p. 101.

Mrs Curd. *Ellen Carrington: A Tale of Hull During the Cholera in 1849*. Hull: Thornton and Pattinson, circa 1850.

G. Eliot. *Daniel Deronda*. Edinburgh: Blackwood, 1876.

M. Bakhtin. *Problems of Dostoevsky's Poetics*. Minneapolis: University of Minnesota Press, translated by C. Emerson, 1984, p. 287.

LINE DRAWINGS

Endpapers: Detail of map of Hull with street names changed, original engraved by Goodwill & Lawson for *Stephenson's Hull Directory*, 1842, Hull Public Library.

Pen and ink bottle: D. Geyer. *Collecting Writing Instruments*. West Chester, PA: Schiffer Publishing, p. 20.

Pleyel Grand Piano: Corbis, 902 Broadway, New York, NY 10010.

Ross Microscope: The Royal Microscopical Society.

Water Cart: M. and C.H.B. Quennell. *A History of Everyday Things in England, 1733–1851*. London: Batsford, Fig 7, p. 7.

Ship: Ocean Steam Navigation Co. *Washington*, 1847. Science Museum, London.